Neurophenomenology and Its Applications to Psychology

Susan Gordon
Editor

Neurophenomenology and Its Applications to Psychology

Springer

Editor
Susan Gordon
Department of Psychology
National University
La Jolla, CA, USA

Southbury Clinic for Traditional Medicines
Southbury, CT, USA

ISBN 978-1-4614-7238-4 ISBN 978-1-4614-7239-1 (eBook)
DOI 10.1007/978-1-4614-7239-1
Springer New York Heidelberg Dordrecht London

Library of Congress Control Number: 2013937293

Printed on acid-free paper

Springer is part of Springer Science+Business Media (www.springer.com)

This book is dedicated to
Francisco J. Varela
(September 7, 1946–May 28, 2001)

and to

Eugene I. Taylor
(October 28, 1946–January 30, 2013)

In Memoriam

Reflections on Eugene Taylor

Two weeks before I presented this manuscript to Springer for publication, Dr. Eugene Taylor passed away. He confided to me his life-threatening illness in February 2012. I must confess that Eugene was always my primary inspiration for this book. He was a beloved teacher, mentor, colleague, and friend during my years at Saybrook Graduate School where he chaired my dissertation and as his research assistant for 9 years at Harvard.

The year before I graduated from Saybrook, Eugene invited me to the American Psychological Association (APA). Five years later, I had the pleasure of presenting him with the 2011 Abraham Maslow Award from Division 32, "for outstanding and lasting contribution to the exploration of the farther reaches of human spirit," a fitting tribute to his character, ethics, and his dedication to existential-humanistic and transpersonal psychology and the Buddhist spiritual tradition.

Upon my graduating from Saybrook, Eugene and I embarked on a writing project to uncover the influence of William James on Francisco Varela, particularly with regard to James' radical empiricism, the subject of his chapter (this volume). I voraciously read Varela's primary works and met with his wife Amy and colleagues in Paris during the summer of 2011. In addition to this project, I assisted Eugene with *William James and the Spiritual Roots of American Pragmatism* and *The Mystery of Personality: A History of Psychodynamic Theories* (Springer, 2009). Eugene's legacy lives on in the hearts and minds of his students and colleagues who dearly loved him and whose lives he so profoundly touched.

Eugene was a paradox. Impassioned and erudite, he was highly emotional, deeply caring, and compassionate. While he evoked fear and awe in many, we were kindred spirits. I recognized his transcendent presence in my life the instant we met. He asked penetrating questions and had a ravenous appetite for precision that rivaled his thirst for knowledge. Our friendship began at a Thai restaurant in Santa Rosa, CA in 2004. As synchronicity would have it, I was seated at a table directly opposite, but facing Eugene, and my husband Andrew and I invited him to join us.

We spoke for hours as we shared what would become numerous conversations about the mind-brain relationship, holistic medicine, Henry Murray, cyberphysiology, Asian and contemplative philosophy, Ellenberger's *The Discovery of the Unconscious* (1970), mystical states, psychical research, James' tripartite metaphysics, self-realization, and Eugene's vision of a person-centered science. I fell in love with Eugene's sheer brilliance, his devotion to living in the moment, higher purpose, veracity, valor, tenderness, sense of humor, and grace.

Eugene held a master's degree in experimental psychology with a minor in Asian studies from Southern Methodist University (SMU) for his thesis "Psychological Suspended Animation: Heart Rate, Blood Pressure, Time Estimation, and Introspective Reports from an Anechoic Environment" (1973) and a Ph.D. in the history and philosophy of psychology from the Boston University Professors Program under the direction of Sigmund Koch for his dissertation, "Psychology as a Person-Centered Science: William James after 1890" (Taylor, 1992a), republished as *William James on Consciousness beyond the Margin* (1996). Eugene also uncovered James' notes for a series of 1896 Lowell lectures at Harvard keyed to marginal annotations in books from James' personal library, which he reconstructed as *William James on Exceptional Mental States* (1982/1984).

Eugene was Executive Faculty and Department Chair in Humanistic and Transpersonal psychology at Saybrook University, a Lecturer on Psychiatry at Harvard Medical School, Senior Psychologist on the Psychiatry Service at the Massachusetts General Hospital, and historian in the Department of Psychiatry for 25 years. He received early training in experimental psychology from psychophysicist William H. Tedford, and in the history of psychology from Jack Roy Strange. Eugene was introduced to the techniques of historical scholarship in comparative religions by Frederick Streng, a student of Mircea Eliade at the University of Chicago, a Buddhist scholar, and specialist in Buddhist-Christian dialogue. Eugene took courses in comparative religions in the Department of Religion at SMU and Perkins School of Theology and applied these techniques to archival investigation in the history of American psychology and psychiatry at Harvard.

He wrote numerous publications on medical biography, the history of medical psychology, and the introduction of psychotherapeutic methods into neurology, psychiatry, psychology, and religion in the United States in the late nineteenth century. From 1977 to 1979, he was a Resident Graduate in the Psychology of Religion and Asian Studies at Harvard Divinity School, where he returned in 1983 as the William James Lecturer on the *Varieties of Religious Experience* (1902). He was the steward of the papers of Gordon W. Allport between 1979 and 1984, where he created the indices for Allport's papers and correspondence, and from 1982 to 1988 he was research assistant to Henry A. Murray. In 1992, he became the founder and director of the Cambridge Institute of Psychology and Religion. Additionally, he authored two books on popular American spirituality, *A Psychology of Spiritual Healing* (1997) and *Shadow Culture: Psychology and Spirituality in America* (1999).

Eugene held memberships in the American Psychological Association (Divisions 1 [Fellow], 24, 26 [Fellow], and 32), American Academy of Religion,

American Association for the History of Medicine, The History of Science Society, The Cheiron Society (International Society for History of the Behavioral and Social Sciences), International Association for the History of European Psychiatry, American Association for the Advancement of Science, International Platform Association, The Thoreau Society, The Cambridge Historical Society, The United States Aikido Federation, and The William James Society, the Philemon Foundation, and was a founding member of the New Existentialists.

Eugene's message to psychology regarding the science of consciousness, psychology as epistemology, the phenomenology of the science-making process, and the humanistic implications of the neuroscience revolution continues to guide my work. It is with great love, appreciation, and gratitude that I dedicate this book to Dr. Eugene Taylor.

References

Ellenberger, H. (1970). *The discovery of the unconscious*. New York, NY: Basic Books.

James, W. (1902). *The varieties of religious experience: A study in human nature*. New York, NY: Longmans, Green and Company.

Taylor, E. I. (1973). *Psychological suspended animation: Heart rate, blood pressure, time estimation, and introspective reports from an anechoic environment*. Master of Arts Thesis, Department of Psychology, Southern Methodist University, Dallas, TX. Privately printed by The Essene Press.

Taylor, E. I. (1982). *William James on exceptional mental states: Reconstruction of the 1896 Lowell lectures*. New York, NY: Charles Scribner's Sons; reproduced in paperback by the University of Massachusetts Press, Amherst, 1984.

Taylor, E. I. (1992a). *Psychology as a person-centered science: William James after 1890*. (Doctoral dissertation, Boston University). *Dissertation Abstracts International, B 53/03*. (UMI No. 9222837).

Taylor, E. I. (1996). *William James on consciousness beyond the margin*. Princeton, NJ: Princeton University Press.

Taylor, E. I. (1997). *A psychology of spiritual healing*. West Chester, PA: Chrysalis Books.

Taylor, E. I. (1999a). *Shadow culture: Psychology and spirituality in America*. Washington, D.C.: Counterpoint.

Taylor, E. I. (2001–2002). *William James and the spiritual roots of American pragmatism*. Lectures commemorating the Centenary of William James's *Varieties of Religious Experience*, delivered at Harvard University. Manuscript completed. Unpublished.

Taylor, E. I. (2009). *The mystery of personality: A history of psychodynamic theories*. New York, NY: Springer.

Foreword

It was bound to happen. Neuroscience is a branch of the life sciences that deals with the anatomy, physiology, biochemistry, or molecular biology of nerve cells and nervous tissue and their relationship with learning and behavior. Cognitive neuroscience quickly became a popular field of study, one that focused on what goes on in the brain when people think, reflect, imagine, consider, reason, explain, and speculate – in other words, when they cognize. Soon a related field developed, that of affective neuroscience with the brain's activity during emotions, feelings, impulses, and moods. Brain scans were applied to other activities as well, for example, neuroeconomics investigates nervous system activity during buying, selling, trading, shopping – and stealing. Computational neuroscience deals with network theory, ranging from brain networks to computer networks. Neurotheology speculates about the brain's capacity for religious, spiritual, and mystical experiences.

At Saybrook University, we have an introductory neuroscience course as well as two courses that apply neuroscientific theory and research to learning disabilities and to sleeping and dreaming. But we also have courses devoted to a very different methodology, that of phenomenology, a philosophical method of inquiry into people's everyday experience. Rather than using a third-person approach, as do the neurosciences, phenomenology takes a first-person approach to the examination of different phenomena as someone becomes aware of them and gives them meaning. Edmund Husserl is regarded as the philosopher whose work made phenomenology a coherent methodology; he felt that awareness is characterized by intentionality because it is always "about something."

And so it was bound to happen. The psychiatrist and neurologist Erwin Straus is generally regarded as the first neurophenomenologist, and I had the privilege of introducing him at a European humanistic psychology conference in the early 1970s. I was transfixed as he eloquently explored the "embodied mind" and criticized mechanistic and reductionistic ways of understanding people and cited Maurice Merleau-Ponty for whom perception was "the point of departure." For Merleau-Ponty, "being in the world" depends first on asking "Where am I?" and only secondarily "Who am I?" A fellow faculty member at Saybrook University, Amedeo Giorgi, introduced Merleau-Ponty and such predecessors as Edmund

Husserl to our students and supervised excellent phenomenological dissertations on such topics as the experience of unexpected recovery from a serious illness to the experience of having dreams that come true.

Neurophenomenology is a scientific endeavor that combines neuroscience and phenomenology to study experience from the perspective of the embodied condition of the human mind. The label was coined by my longtime colleague Charles Laughlin and his associates in the early 1990s and was appropriated by Francisco Varela who wrote some of the defining papers and books in this area. Varela's work was seminal to one of my students, Susan Gordon. I served on her dissertation committee and was the discussant for her pioneering symposium at the American Psychological Association's 2009 convention, a symposium titled "Neurophenomenology and the Enactive Approach to Cognition," a forum that served as the nexus for this book. Dr. Gordon's work is relevant for psychologists across the professional spectrum and I believe that it has important implications for the future of psychological science.

Historically, phenomenological psychology has not had the same level of scholarly debate that it once had in philosophy. However, this book introduces neurophenomenology and its relation to the natural sciences. It calls attention to the philosophical tensions in these areas so that they can become more accessible to psychologists. Neurophenomenology provides a reinterpretation of both cognitive and affective neuroscience within the theoretical framework of phenomenology and a bridge between first-, second-, and third-person perspectives in the study of experience.

Dr. Gordon points out how cognitive behaviorism, a reductionistic and positivistic discipline, and phenomenology, a descriptive psychology of immediate experience, provide two contrasting epistemological bases for defining and establishing psychology as a science. As an empirical discipline, neurophenomenology can help clarify psychology's relationship to philosophy because it introduces an approach to philosophy and human science that is able to articulate the reality-based ground upon which science itself operates. It can also help the neurosciences to understand the meaning and metaphysics of lived experience as well as reexamine scientific ways of knowing.

This book introduces psychologists to the field of neurophenomenology as a way to integrate interdisciplinary knowledge with the fields of psychological theory, research, and practice. Most books on neurophenomenology are written for philosophers. There are few, if any, books on this topic, one that reinterprets neurophenomenology in a psychological context. To the best of my knowledge, this is the first to rise to the challenge. With regard to the field of neurophenomenology in general, this book has important contributions to make for the development of the field because it challenges the dominant reductionistic worldview that governs the definitions of problems and methodologies in both the physical and the life sciences.

This book is composed of five provocative chapters: Chapter 1 explores the neurophenomenology of emotion, Chapter 2 applies neurophenomenology to learning and thinking, Chapter 3 describes the mapping of attention and awareness in Tibetan meditators, Chapter 4 traces the historical and theoretical basis of William James' philosophy of radical empiricism for present-day neuroscience, and Chapter 5 constructs a model of the embodied, experiencing neurophenomenological self. All of these chapters delineate the applications of neurophenomenology to psychology emphasizing the role it can play in humanizing science, arguing for an embodied approach to experience. Presently, this approach is unknown to many branches of psychology, even though existential-phenomenological and humanistic-transpersonal perspectives have provided this perspective for decades.

Dr. Gordon's model is eclectic and interdisciplinary. It reveals the mind-brain interface, the growth-oriented dimension of the person, and the myth-making dimension of human experience. My colleagues and I were aware of this process when we wrote about "personal myths" or existential life beliefs, stressing their roots in an individual's biochemistry.

Dr. Gordon describes her model as "psychoneurointracrine" because it relates psychological constructs and neurological activity to intracrine, intracellular processes, in other words, the biosynthesis of steroids within cells, the binding of receptors, and enzyme formation that catalyzes the creation of intracellular hormones. Dr. Gordon's model is growth oriented and self-organizing, leading to what she calls the "neurophenomenological self." Her model has considerable explanatory value, providing a way to bridge the gap between mind and brain, how this mind-brain develops not only meaning but one's sense of well-being, and – ultimately – how the notion of "self" emerges from this complex system. Unlike other interdisciplinary approaches such as psychoneuroimmunology that focus on illness, Dr. Gordon's psychoneurointracrinology is growth oriented, bringing it under the rubric of humanistic, existential, and transpersonal psychology. Indeed her work may serve as a stimulus to a renewed vigor in psychology as it evokes a model that integrates a unique philosophical school, phenomenology, with cutting-edge developments in the neurosciences.

Professor of Psychology and Integrative Inquiry
Saybrook University
San Francisco, CA, USA

Stanley Krippner, Ph.D.

Editor's Preface

This book began with a symposium titled "Neurophenomenology and the Enactive Approach to Cognition," sponsored by the Societies for Humanistic Psychology (Division 32) and Theoretical and Philosophical Psychology (Division 24), which I chaired at the 117th annual convention of the American Psychological Association in Toronto, Canada, in 2009. This symposium presented the collection of papers featured in this volume by five psychologists who aspire to interpret neurophenomenology to psychologists: reenvision the place of phenomenology within psychological science and the sciences in general; and define the contributions of existential-phenomenological and humanistic-transpersonally oriented psychology to this discussion.

Existential-phenomenological and humanistic-transpersonal perspectives focus on what it means to be fully, experientially human. They are concerned with the individual's creation of meaning, actualization of values, and potential for self-realization. Figures such as William James, Abraham Maslow, Carl Rogers, Rollo May, Gordon Allport, Lois and Gardner Murphy, Paul Tillich, Karen Horney, Erik Erickson, Carl Jung, Erich Fromm, Jean Paul Sartre, Martin Buber, and others focused their writings on the self that is directly experienced, on the actualization of potential, on striving toward health as intrinsic to human motivation, and on existential themes inherent to interior exploration. Their vision went beyond the measurement of behavior to embrace a wider view of personality than mainstream trait theories because they acknowledged a growth-oriented dimension of the person.

From 1940 to 1970, humanistic psychology, centered on transforming reductionist experimentalism, pioneered the emergence of a person-centered, growth-oriented, existential psychology of the whole person. It advanced a dialogue between science and the humanities within the Western university system and flourished as a viable form of academic discourse (Taylor, 1999, 2009). The lineage of humanistic psychology spans the person-centered science and psychology of William James in the 1890s and early 1900s; the macropersonality theories and social psychologies of Gordon Allport, Henry Murray, and Gardner Murphy in the 1930s and 1940s; and the self-actualizing and motivational psychologies of Carl Rogers and Abraham Maslow and the European existential-phenomenological psychotherapeutic traditions united by Rollo May and Henri Ellenberger in the 1950s and 1960s.[1] As Taylor notes, humanistic psychology, as an academic discourse arising out of personality theory and motivational psychology, was absorbed by the American psychotherapeutic counterculture and split into three streams by the late 1960s: transpersonal interest in meditation and altered states of consciousness, experiential encounter groups and somatic bodywork, and human science and radical political psychology. Transpersonal psychology, which developed from the humanistic movement after 1969, began through the experiential study of entheogens, meditation, altered states of consciousness, and non-Western epistemologies.[2]

Phenomenology and neurophenomenology introduce an approach to philosophy, psychology, and human science that is able to articulate the ontological ground upon which science itself operates in ways that are not accessible to positivist science. Cognitive behaviorism, the present standard in mainstream theoretical and clinical psychology, founded on reductionistic positivism, can neither address the implications of the neuroscience revolution regarding the phenomenology of consciousness nor the relationship between the mind and the brain, because it lacks a self-reflective and prereflective element. Reliability, validity, standardization, prediction, and control are all subsets of holism and qualitative experience. Science ignores human consciousness in operationally defining the person when it disregards phenomenological data that it cannot see or measure. Science needs grounding in self-reflection, contemplation, and an embodied approach to experience that is unrestricted to the study of behavior and a phenomenologically oriented psychology that is foundational to the sciences.

Phenomenology studies consciousness, both the rational waking state as well as dynamics of the unconscious, as it is experienced from the first-person point of view. Experience must be grasped holistically as a relationship in which the subject relates to an object through its meaning. Classical approaches in phenomenological practice have ranged from the reflective analysis of lived experience as it presents itself (Husserl, Merleau-Ponty) to contextual hermeneutic phenomenology (Heidegger), radical empiricism (James), logico-semantic models that analyze the form of experience or specify conditions of truth, and empirical experiments that confirm or refute aspects of experience. On the basis of Husserl's epistemology, Heidegger, Sartre, and Merleau-Ponty pioneered phenomenological studies of existence (phenomenological ontology) diverging from Husserl's phenomenological reduction (PhR), which provided an intuitive method to transcend the natural attitude. Phenomenological methods allow observation of internal states of consciousness using a meditative focus to loosen presumptions, so that we may begin to understand the essence of lived experience. Neurophenomenology has attempted to *naturalize phenomenology* (Petitot, Varela, Pachoud, & Roy, 1999) training neuroscientists to become reflectively aware of the structure of experience in the conduct of research (Petitmengin, 2009; Varela & Shear, 1999, Zahavi, 2008).

However, while many recent books in neurophenomenology address the relationship between the mind and the brain and the nature and structure of consciousness, this pioneering work is largely inaccessible to psychologists who do not have a framework for applying it to their discipline. I would argue that this is because neurophenomenology requires translation from the language of phenomenology, cognitive science, and non-Western epistemology to that of psychological science. A case on point is Shaun Gallagher and Dan Zahavi's *The Phenomenological Mind: An Introduction to Philosophy of Mind and Cognitive Science* (2008).

For example, in his review of their book, published in the *Journal of Phenomenological Psychology*, Amedeo Giorgi (2009b), author of *The Descriptive Phenomenological Method in Psychology* (2009a), states:

> As I read this book, a certain envy overcame me. If only, in psychology we had such a dialogue going between phenomenological and mainstream, empirical approaches to psychological phenomena. However, in psychology, not only is phenomenology mostly ignored, even where it does occasionally show up, it is poorly understood. A possible secondary outcome of the book under review is that such a dialogue could trickle down to the psychological level. (Giorgi, 2009b, p. 108)

Giorgi's five-step research method uses Husserlian phenomenology as its philosophical foundation.[3] However, because Husserl's work was written for philosophy, Giorgi (2009a) needed to adapt the principles of his phenomenological reduction for psychological investigation. Giorgi's (1970) method provides the systematic rigor of empirical science but is not reductionistic in its treatment of the person. Subjects describe the structure of psychological phenomena so that it can be understood in a deeper, holistic, and more comprehensive way. It is the meaning of experience, as it is lived in the body (embodied), rather than the objective interpretation of behavior that is essential to phenomenology.

The authors of this volume introduce neurophenomenology to suggest steps toward a more experiential, nonreductive, phenomenologically oriented, descriptive, person-centered psychology of immediate experience. Neurophenomenology introduces a theoretical and practical framework that integrates the natural and human sciences to consciousness, which invites an interdisciplinary dialogue on the nature of awareness, the ontological primacy of experience, the perception of the observer, and the mind-brain relationship that will shape the future of psychological theory, research, and practice.

What is neurophenomenology? This term, coined by Laughlin, McManus, and d'Aquili (1990), was distinguished as a new research direction for the neuroscience of consciousness by Francisco Varela and colleagues in the mid-1990s. Neurophenomenology bridged ideas from systems theory, cognitive computationalism, and autopoiesis by combining first- and third-person methods in experimental research. *First-person methods* refer to phenomenological lived experience, the contemplative study of attention, present-time consciousness, body image, volition, perception, intentionality, fringe, centre, and emotion associated with subjective mental states. *Third-person methods* refer to the analysis of neurophysiological data from the measurement of large-scale sensorimotor processes in the brain using fMRI, EEG, MEG, and cognitive testing. *Second-person perspectives*, the empathic, intersubjective, interpersonal dimensions of conscious experience, are also investigated using phenomenological studies that borrow primarily from non-Western epistemology and the work in philosophy by Edmund Husserl, Martin Heidegger, and Maurice Merleau-Ponty (e.g., Varela & Shear, 1999; Petitmengin, 2009; Varela, Lachaux, Rodriguez, & Martinerie, 2001; Thompson, 2001a).

This book explores the meaning and import of neurophenomenology (i.e., phenomenology in its current relation to the natural scientific studies of the nervous system), the philosophy of *enactive* or embodied cognition, and the theory of autopoiesis interpreted for psychologists. *Embodiment* refers to the bodily aspects of human subjectivity: the biological and physical presence of our body as a necessary precondition for the experience of emotion, language, thought, and social interaction. It provides a systematic and dynamical framework for understanding how a cognitive self – a mind – can arise in an organism in the midst of its operational cycles of internal regulation and outgoing sensorimotor coupling (Rudrauf, Lutz, Cosmelli, Lachaux, & Le Van Quyen, 2003). *Autopoiesis* explains the continuity of mind and life observed in the self-organizing properties of chemical, neuronal, and cognitive systems through which they continually regenerate, recreating themselves by their own mutual interactions.

Varela and colleagues have explored ontogenic developmental learning, perception-action in the synchronous coupling of neuronal cell assemblies, and present-time consciousness, philosophically rooted in the functionalism of William James and the European clinical traditions of Edmund Husserl, Martin Heidegger, Maurice Merleau-Ponty, and Jean Paul Sartre in existential-phenomenology (Varela, Thompson, & Rosch, 1991). Neurophenomenology has influenced the melding of traditional boundaries between continental and analytical thought, the move from behaviorism to cognitive neuroscience, and the affective and experiential revolutions presently underway in psychology. Instead of viewing the mind as an epiphenomenon of the brain, consciousness is seen as a distributed phenomenon of the whole active organism. Mental life is situated in the world and consciousness is intersubjectively enacted in interdependency with its surroundings through action, perception, emotion, and the self-moving flow of time consciousness (Thompson, 2007) from which meaning becomes inseparable.

Rooted in Varela's knowledge of biological systems and Indo-Tibetan Buddhism, a central question for neurophenomenology has been how consciousness and subjective experience relate to the brain and the body. While the "easy problems" for neuroscience pertain to questions about the difference between wakefulness and sleep and the mechanisms that allow us to focus our attention, the "hard problem" is the relationship between objective knowledge and subjective experience or *qualia*, the ineffable conscious experience, as distinct from the physical or computational process of the brain. Giving a fundamental role to first-person accounts and the irreducible nature of experience, neurophenomenologists address the problem of consciousness by establishing heuristic mutual constraints between biophysical data (third person) and the data produced by accounts of subjective experience (first person) as a co-emergent, enactive process (Lutz & Thompson, 2003; Petitot, Varela, Pachoud, & Roy, 1999; Varela, 1996).

In her tribute to Varela, Petitmengin (2009) points to the prereflective and implicit character of lived experience viewed from within as the most immediate and intimate thing about us that is not directly accessible but requires a method. She notes that becoming aware of prereflective experience is not a process of distancing and objectification or a fracturing of the self between an observer and an observed. In psychological context, it means coming into closer contact with one's experience, not via accumulating new knowledge but by striping ourselves of the knowledge that prevents us from entering into contact with our true nature and pure experience. Neurophenomenology is that method with its many applications to psychology.

About This Book

Each of the five chapters in this book cohesively contributes to psychology's understanding of neurophenomenology. I have situated each to build on the foundations of the previous chapters.

In Chapter 1, "Enactive Cognition and the Neurophenomenology of Emotion," **Brent Robbins** explores theories of emotion from a perspective that integrates neurobiological, cognitive, and phenomenological approaches to consciousness and examines how this synthetic approach may resolve current difficulties in the conceptualization of emotion in psychology. Robbins argues that the enactive approach calls into question an old paradigm of the theory of emotion, which conceptualizes emotion and cognition as distinct functions located in separate regions of the brain. An account of emotion, instead, needs to preserve the meaning of the experience as it appears within the life-world context of the person, rather than being based on inferences drawn from laboratory conditions. He illustrates how we need to step back to examine how form is constituted as an object for cognition; how cognition, emotion, and perception appear to be well-integrated processes that cannot be teased apart without making artificial distinctions; and how the complex web of emotional experience requires a first-person perspective. Robbins argues that enactive and neurophenomenological approaches are promising avenues for bringing forth an affective, experiential revolution in psychology to fruition.

In Chapter 2, "Neurophenomenological Praxis: Its Applications to Learning and Pedagogy," **Robert McInerney** introduces neurophenomenology to the psychologist-educator and illustrates a prereflective, situated, enactive assessment of learning and thinking using concepts from autopoiesis and neurophenomenology. This chapter details the phenomenological approach and provides specific examples to illustrate enactive, embodied learning. McInerney discusses the theoretical basis of mind-body dualism, the perspectives of cognitivism and Husserl's phenomenological method, as well as the phenomenologies of James and Dewey in a psychological and pedagogical analysis of learning inherent to philosophy and epistemology. He outlines how neurophenomenological praxis can lead to a pedagogy that recognizes and liberates essential forms of learning that have been devalued by our educational system as well as the practical applications of this method to portfolio learning and assessment.

In Chapter 3, "Mutual Enlightenment: Cognitive Phenomenology in the Study of Tibetan Meditation," **Olga Louchakova-Schwartz** presents original neurophenomenological research and discusses its theoretical, empirical, and practical applications to the study of cognition and the nature of prereflective awareness in meditation. She illustrates this relationship through an analysis of the types of meditation and discusses her method of *phenomenological-cognitive mapping* for studying cognitive changes during meditation. This method helps to translate the empirical conditions of meditation into cognitive psychological experimental research. Louchakova-Schwartz reports on the experimental results of enhancement of visual imagery in Tibetan meditators, discusses her neurophenomenological approach, and presents a comparative phenomenological analysis of four styles of meditation in Tibetan Buddhism (Rig-pa, Vipashyana, Mandala, and Deity). She argues that the phenomenology is crucial for a successful experimental design.

In Chapter 4, "Déjà-Vu: William James on 'The Brain and the Mind,' 1878," **Eugene Taylor** explores the historical and theoretical basis of James' philosophy of radical empiricism for present-day neuroscience through a discussion of James' Lowell Lectures of 1878 on "The Brain and the Mind," as a way to remind us that the problems of neuroscience today were broached more than 125 years ago. Taylor explains how Varela and his interpreters are on the cusp of a breakthrough in understanding James' philosophy of radical empiricism and the role of the intersubjective observer for a person-centered approach to science. He argues that the implications of this breakthrough have the potential to address the so-called hard problem, namely, the relation between the brain and the mind, which could possibly set the stage for an examination of the phenomenology of the science-making process itself. Such a new science would account for the *weltanschauung* of the experimenter, the intersubjective relation between the observer and the observed, and alter our understanding of the presence of the experimenter on the outcome of what he or she studies.

In Chapter 5, "Psycho-neuro-intracrinology: The Embodied Self" (**Susan Gordon**), I introduce an autopoietic model of the *neurophenomenological self* or growth-oriented dimension of the person as the confluence of psychological, neurological, and intracrinological systems. Two theories are advanced to explain how the self has correlates not only in the brain, but also in the connections between the hypothalamic-pituitary-gonadal and hypothalamic-pituitary-adrenal (HPG-HPA) axes, which are responsible for enactive engagement and the development of meaning through their connections to the higher-order functions of the brain. The theory of psychoneurointracrine autopoiesis explains how the regulation of a steroid's receptor is modulated by the person's perception of experience and sense of well-being. The theory of emergent global states examines how reciprocal limbic projections from the HPG-HPA axes integrate prereflective, autonomic, subliminal, and archetypal experience in the development of meaning and the emergence of the self. These theories extend knowledge of the mind-brain relationship and the growth-oriented dimension of the person.

These chapters proceed from the neurophenomenology of emotion to an analysis of neurophenomenological praxis in learning and pedagogy, to a neurophenomenological study of visuospatial process in Tibetan meditators, to a theoretical and philosophical commentary on the relations between neurophenomenology, radical empiricism, and the future of scientific psychology, to a psychoneurointracrine model of the embodied, neurophenomenological self. Emotion is explored as the ground from which cognition occurs, and prereflective awareness is examined as foundational to the lived experience of meaning and the growth-oriented dimension of the person. The ideas presented in this volume have wide application to psychological science: understanding the experience of emotion, expanding our methods of teaching and learning, the value of research on meditation to an understanding of consciousness, the implications of James' epistemology for present-day neuroscience, and an embodied approach to experience. By integrating Western Anglo-American and Continental phenomenology with cognitive science and Eastern contemplative experience and practices, neurophenomenology provides a bridge between the sciences that neither reduces the mind to the physiology of the brain nor the living organism to cause and effect relationships, but instead provides steps toward a more person-centered science. It is my hope that this book stimulates a rich and fruitful academic discourse for psychologists across the professional spectrum as well as interdisciplinary scholars of phenomenology, neuroscience, philosophy, and consciousness.

References

Buytendijk, F. J. J. (1967). Husserl's phenomenology and its significance for contemporary psychology. In N. Lawrence & D. O'Connor (Eds.), *Readings in existential-phenomenology*. Englewood Cliffs, NJ: Prentice Hall.

Churchill, S. D. (2010). "Second person" perspectivity in observing and understanding emotional expression. In L. Embree, M. Barber, & T. J. Nenon (Eds.), *Phenomenology/Selected essays from North America, Part 2: Phenomenology beyond Philosophy* (Vol. 5). Bucharest/Paris: Zeta Book/Arghos-Diffusion.

Gallagher, S., & Zahavi, D. (2008). *The phenomenological mind: An introduction to philosophy of mind and cognitive science*. New York/Abingdon/Oxon: Routledge.

Giorgi, A. (1970). *Psychology as a human science*. New York: Harper and Row.

Giorgi, A. (2009a). *The descriptive phenomenological method in psychology*. Pittsburgh, PA: Duquesne University Press.

Giorgi, A. (2009b, May, 1). [Review of the book *The phenomenological* mind: An introduction to philosophy of mind and cognitive science, by S. Gallagher and D. Zahavi]. *Journal of Phenomenological Psychology, 40*(1), 107–125.

Giorgi, A. (2010). Phenomenological psychology: A brief history and its challenges. *Journal of Phenomenological Psychology, 41*, 145–179.

Gurwitsch, A. (1966). The place of psychology in the system of sciences. In A. Gurwitsch (Ed.), *Studies in phenomenology and psychology*. Evanston, IL: Northwestern University Press.

Hardy, L. (Trans.) (1990). *Edmund Husserl: The idea of phenomenology: A translation of Die Idee der Phänomenologie Husserliana II*. Dordrecht, Boston, London: Kluwer Academic Publishers.

Laughlin, C., McManus, J., & d'Aquili, E. (1990). *Brain, symbol and experience: Toward a neurophenomenology of consciousness*. New York, NY: Columbia University Press.

Lutz, A., & Thompson, E. (2003). Neurophenomenology: Integrating subjective experience and brain dynamics in the neuroscience of consciousness. *Journal of Consciousness Studies, 10*(9–10), 31–52.

Merleau-Ponty, M. (1962). *Preface in Phenomenology of perception* (pp. vii–xx). New York, NY: Humanities Press.

Petitmengin, C. (Ed.). (2009). *Ten years of viewing from within: The legacy of Francisco Varela*. Thorverton, UK: Imprint Academic.

Petitot, J., Varela, F. J., Pachoud, B., & Roy, J.-M. (Eds.). (1999). *Naturalizing phenomenology: Issues in contemporary phenomenology and cognitive science*. Stanford, CA: Stanford University Press.

Rudrauf, D., Lutz, A., Cosmelli, D., Lachaux, J.-P., & Le Van Quyen, M. (2003). From autopoiesis to neurophenomenology: Francisco Varela's exploration of the biophysics of being. *Biological Research, 36*, 27–65.
Strasser, S. (1967). Phenomenologies and psychologies. In N. Lawrence & D. O'Connor (Eds.), *Reading in existential-phenomenology*. Englewood Cliffs, NJ: Prentice-Hall.
Taylor, E. (1999). *Shadow culture: Psychology and spirituality in America*. Washington, D.C.: Counterpoint.
Taylor, E. (2009). *The mystery of personality: A history of psychodynamic theories*. New York, NY: Springer.
Thompson, E. (Ed.). (2001). *Between ourselves: Second-person issues in the study of consciousness*. Charlottesville, VA: Imprint Academic.
Thompson, E. (2007). *Mind in life: Biology, phenomenology, and the sciences of mind*. Cambridge, MA: Harvard University Press.
Varela, F. J. (1996). Neurophenomenology: A methodological remedy for the hard problem. *Journal of Consciousness Studies, 3*(4), 330–349.
Varela, F. J., & Shear, J. (Eds.). (1999). *The view from within: First-person approaches to the study of consciousness*. Thorverton, UK: Imprint Academic.
Varela, F. J., Lachaux, J. P., Rodriguez, E., & Martinerie, J. (2001). The brainweb: Phase synchronization and large-scale integration. *National Review of Neuroscience, 2*(4), 229–239.
Varela, F. J., Thompson, E., & Rosch, E. (1991). *The embodied mind: Cognitive science and human experience*. Cambridge, MA: The MIT Press.
Zahavi, D. (2005). *Subjectivity and selfhood: Investigating the first-person perspective*. Cambridge, MA: The MIT Press.

Notes

1. Rollo May, Henri Ellenberger, and others became central figures uniting the separate European traditions of existentialism and phenomenology under the umbrella of humanistic psychology in the form of existential-phenomenological psychotherapy. Others such as Charlotte Bühler, James Bugental, Adrian van Kaam, and Sydney Jourard wrote on humanistic themes in existential psychology. It was Rogers, Maslow, and May; however, who established a new norm for psychology as a whole, despite opposition from behaviorists and psychoanalysts, declaring that humanistic psychology, at the center of their vision of a transformed discipline, was person centered, growth oriented, and existential in orientation (Taylor, 2009, pp. 263–264).
2. Figures out in the wider culture, such as Alan Watts, a student of Zen teachings and Episcopal minister, his teacher D. T. Suzuki, the theosophist Jiddhu Krishnamurti, Indian yogis such as Swami Rama, psychophysiologists such as Elmer and Alyce Green, indologists and religious philosophers such as Frederic Spiegelberg and Huston Smith, and Vedantic practitioners such as Aldous Huxley and Gerald Heard inoculated Westerners with concepts of consciousness and techniques of meditation drawn from classical Asian psychology and other world religions. This was also the time when psychedelic drugs were first introduced into the general population and had the effect on the resurgence of a popular spiritual psychology (Taylor, 2009, p. 264).
3. For an introduction to the phenomenological approach to psychology taught by Giorgi and others, see Buytendijk (1967), Giorgi (2010), Gurwitsch (1966), Merleau-Ponty (1962), and Strasser (1967). *Note: Relational* intentionality, which is what empirically grounded, phenomenological psychologists do within the natural attitude (Churchill, 2010), is different from *adverbial* intentionality, which refers to Husserl's transcendental philosophy that focuses on the process of consciousness "itself" as bracketed from the "transcendent reality" that surrounds it (Hardy, 1990).

Acknowledgments

I offer heartfelt thanks to my husband and family for their encouragement, support, and love throughout the writing of this book.

Formal acknowledgements are gratefully extended to Jean Lassègue, Director of Centre de Recherche en Epistémologie Appliquée (CREA), École Polytechnique, for his introduction to Michel Bitbol, Director of Research at Centre National de la Recherche Scientifique (CNRS); and to Michel Bitbol for his introduction to David Chavalarias, Co-director, L'Institut des Systèmes Complexes and to Amy Cohen Varela.

Acknowledgements also to William Irwin Thompson and Elaine Engst, Archivist at Cornell University, for permission and access to Dr. Thompson's papers at the Kroch Library, Division of Rare Books and Manuscripts; particularly Lindisfarne Newsletters, early papers of Francisco Varela, photos, and personal correspondence.

I am indebted to the Widener Library at Harvard and the Bibliothèque François Mitterrand in Paris for their assistance with my research. I extend special thanks to Dorothy Barr, Public Services Librarian in Molecular & Cellular Biology at Harvard's Ernst Mayr Library for sending me Francisco Varela's dissertation; and to Debra Case Lillian for her help translating Varela's works written in French.

Sincere thanks is extended to Sharon Panulla, Executive Editor at Springer and her assistant Sylvana Ruggirello, to the contributing authors of this volume for their dedication, and to Stanley Krippner for his guidance. Finally, I offer deepest gratitude to Eugene Taylor for his inspiration, devotion, and vision throughout our many years of friendship.

Contents

Editor's Introduction xxix

Enactive Cognition and the Neurophenomenology of Emotion 1
Brent Dean Robbins

Neurophenomenological Praxis: Its Applications to Learning and Pedagogy 25
Robert Garfield McInerney

Cognitive Phenomenology in the Study of Tibetan Meditation: Phenomenological Descriptions Versus Meditation Styles 61
Olga Louchakova-Schwartz

***Déjà Vu:* William James on "The Brain and the Mind," 1878 – A Comment on Current Trends in Neurophenomenology Defining the Application of James's Radical Empiricism to Psychology** 89
Eugene Taylor

Psychoneurointracrinology: The Embodied Self 115
Susan Gordon

Glossary 149

About the Editor and Authors 157

Bibliography 161

Index 187

Contributors

Brent Dean Robbins Point Park University, Pittsburgh, PA, USA

Robert Garfield McInerney Point Park University, Pittsburgh, PA, USA

Susan Gordon National University, La Jolla, CA, USA

Southbury Clinic for Traditional Medicines, Southbury, CT, USA

Olga Louchakova-Schwartz Sofia University, Palo Alto, CA, USA

Eugene Taylor Saybrook University, San Francisco, CA, USA

Harvard Medical School, Boston, MA, USA

Psychiatry Service, Massachusetts General Hospital, Boston, MA, USA

Editor's Introduction

In tribute to the memory of Francisco Varela, one of the chief proponents of the movement in neurophenomenology, I offer my interpretation of his lineage in existential-phenomenology and humanistic-transpersonal psychology, including an account of his existential crisis in the 1970s that altered his epistemological view of science, which inspired this book.

Embodied Mind: Opening an Interdisciplinary Dialogue

The inspiration for our symposium "Neurophenomenology and the Enactive Approach to Cognition" was *The Embodied Mind: Cognitive Science and Human Experience* (Varela, Thompson, & Rosch, 1991) and its sequel *Mind in Life: Biology, Phenomenology, and the Sciences of Mind* (Thompson, 2007), which Thompson and Varela began writing together in 1994. Their approach was to enlarge and enrich philosophical and scientific inquiry and to open an interdisciplinary dialogue: "to bring the experimental sciences of life and mind into a closer and more harmonious relationship with phenomenological investigations of experience and subjectivity" (2007, p. ix) and "to make headway on the most outstanding philosophical and scientific problems of our time – the so-called explanatory gap between consciousness and nature" (p. x). I am in complete agreement with Thompson,

> To make real progress on the explanatory gap, we need richer phenomenological accounts of the structure of experience, and we need scientific accounts of mind and life informed by these phenomenological accounts. Phenomenology in turn needs to be informed by psychology, neuroscience, and biology. (2007, p. x)

Embodied Mind was heavily influenced by the period during the late 1970s when Francisco Varela taught at the summer Science Program of the Naropa Institute in Boulder, Colorado. The goal of Naropa Institute was to create an intellectual space for dialogue between cognitive science and the Buddhist traditions of meditative psychology and philosophy guided by the writings and research of Maurice

Merleau-Ponty, who believed that the embodiment of knowledge encompassed the body as both a lived, experiential structure and as the context or milieu of cognitive mechanisms (1991, p. xvi).

Embodied Mind was also written in the wake of Varela's existential crisis on fleeing his home in Chile after the military coup and overthrow of the Marxist Allende government. This experience brought him to an existential realization that the cognizing subject is a fundamentally fragmented, egoless, groundless, or *selfless self* whose evolution is not optimal adaptation, but a natural drift. His position, as I understand it, is explained below.

Varela designated *neurophenomenology* "a quest to marry modern cognitive science with a disciplined approach to human experience" (1996, p. 330). While he placed himself in the lineage of the Continental tradition of phenomenology and sought to develop a science of consciousness, he claimed not to have ascribed his position on phenomenology to any particular school or sub-lineage, but to his own synthesis in light of modern cognitive science and non-Western traditions focusing on human experience. Varela described phenomenology as "a special type of reflection or attitude about our capacity for being conscious" (1996, p. 335).

Varela believed that neither empirical correlates nor purely theoretical principles were useful to understand *qualia*, the ineffable conscious experience. But, instead, we need to turn our attention to a systematic exploration of the only link between mind and consciousness that seems obvious and natural, *the structure of human experience itself*. Instead of finding extra ingredients to account for how consciousness emerges from matter and the brain, Varela reframed the question to finding meaningful bridges between these two irreducible phenomenal domains:

> What is needed, are precisely the connecting structures provided by Phenomenological Reduction (PhR) since they are both immediately pertinent for experience (by their very nature) and at the same time, sufficiently intersubjective to serve as constructive counterparts for external analysis. (1996, p. 341)

As evident from Varela's writings (1992/1999; 1997, Varela & Shear, 1999) (Gordon, 2009), he was strongly influenced by Husserl's (1893–1917/1991, 1913/1962; 1925/1977, 1893–1917/1950–1966) position on temporality, consciousness, perception, intentionality, and intersubjectivity as well as those of Merleau-Ponty (1962/1945; 1964), James (1890, likely 1912), and Heidegger (1927/1962). According to Petitmengin (2009), whose doctoral work under Varela focused on prereflective and intuitive experience, he was also influenced by philosopher-psychotherapist Eugene Gendlin and psychiatrist-psychoanalytic theorist Daniel Stern, which, from my perspective, situate him squarely in the intellectual lineage of the existential-humanistic tradition in psychology. Gendlin's method of *Focusing* (1978a) emerged from his collaboration with humanistic psychologist Carl Rogers. Daniel Stern's methods examined prereflective experience (which does not recognize itself, is egoless) in the infant's expression of meaning and self-constitution. In addition to his existential-humanistic roots and focus on the non-Western epistemology of Indo-Tibetan Buddhism, Varela's concepts of embodiment, autopoiesis, and self/no-self have their roots in immunology (Vaz & Varela, 1978; Amy Cohen Varela, personal communication, July 9, 2011; Varela, 1979a, 1999) as well as

Gestalt psychology, holistic biology, and anthropology through his associations with Margaret Mead and Gregory Bateson.[1] These relationships are further explored in the following tribute. See also Robbins and Gordon (in press) for an historical perspective.

In Tribute to Francisco Varela

A most heartfelt obituary for Varela (September 7, 1946–May 28, 2001), written by Evan Thompson, published in *PSYCHE* and featured in the documentary *Monte Grande* provides a rich synthesis of his life and works. According to Thompson (2001b), Varela received a classical education from the German Lyceum in Santiago, Chile, where he developed an appreciation for literature, art, philosophy, and science. He earned an M.Sc. in Biology in 1967 from the University of Chile where he studied with neurobiologist Humberto Maturana. Thereafter, he graduated from Harvard with a doctoral degree in Biology at the age of 23 for his dissertation on the phenomenology of perception titled "Insect retinas: Information processing in the compound eye" (1970), written under the direction of Nobel Prize recipient Torsten Wiesel. Why do I think that Varela's fascination with information processing in the visual system was to observe the phenomenology of perception? Varela believed that there is no objective world independent of the observer. As Merleau-Ponty explains in *Phenomenology of Perception* (1962), the theory of the body is already a theory of perception, and "sensation" a unit of the field of experience.

> My field of perception is constantly filled with a play of colours, noises and fleeting tactile sensations which I cannot relate precisely to the context of my clearly perceived world (xi).... Perception is not a science of the world, it is not even an act, a deliberate taking up of a position; it is the background from which all acts stand out, and is presupposed by them. The world is not an object such that I have in my possession the law of its making; it is the natural setting of, and field for, all my thoughts and all my explicit perceptions. (1962, p. xii)

Merleau-Ponty contests the idea that perception is a process by which the external world is imprinted on the subject. Instead, perception is an embodied behavior that emanates from nature.

Varela's dissertation demonstrated the mechanism of visual processing through investigation of color, pattern, and polarized light in the compound eye of the honeybee, as "embodied in its functional organization" (1970, p. 12), which foreshadowed his theory of autopoiesis and the information processing of neuronal networks, synaptic connections, and retinal receptors. As Varela explains in *Embodied Mind* (1991) and *Ethical Know-How* (1992/1999), "different aspects of vision are emergent properties of concurrent subnetworks, which have a degree of independence and anatomical separability, but cross-correlate and work together so that a visual percept is this coherency" (1992, p. 48). He continues,

> Color is not a property that is to be 'recovered' from environmental 'inputs,' in some way. Color is a dimension that shows up only in the phylogenic dialogue between an environment

> and the history of an active autonomous self that partly defines what counts as an environment. Light and reflectances provide a mode of coupling, a perturbation that triggers, that provides an occasion for, the enormous informative capacity of neural networks to constitute sensorimotor correlations and hence put into action their capacity for imagining and presenting. (1992, p. 57)

From 1970 to 1973, Maturana and Varela (1973, 1974, 1980; Varela & Maturana, 1973) formulated their theory of autopoiesis at the University of Chile. According to this theory, living systems are autonomous systems (endogenously controlled, self-organizing, and self-producing) in the form of an operationally closed, membrane-bounded, reaction network. Autopoiesis defined cognition in its minimal biological form as the "sense-making" capacity of life. What they discovered, as enumerated above, was that the nervous system is not an input–output information processing system, but rather an autonomous, operationally closed network whose functional elements are invariant patterns of activity in neuronal ensembles (Varela, 1979a).

Per Thompson (2001b), when Varela returned to Chile, he arrived on September 2, 1970, 2 days before the election of Salvador Allende (the first Marxist politician ever elected in a free election). Three years later, Chile was in turmoil, and Francisco, a strong supporter of the Allende government, was forced to flee with his family after the military coup of General Augusto Pinochet overthrew the government on September 11, 1973. He fled to Costa Rica and, eventually, to the United States, where he took up a position as Assistant Professor at the University of Colorado Medical School and taught and pursued his research until 1978. From my reading of his personal writings (Varela, 1979b), the period in his life from 1970 to 1978 marked the beginning of his existential crisis.

From the mid-1970s, Varela was an ardent practitioner of Indo-Tibetan Buddhist meditation and a student of Buddhist psychology and philosophy. He learned to recognize the emptiness of self and its groundlessness through the practices of mindfulness awareness and *sunyata* (i.e., egolessness, nonduality). According to Thompson (2001b), Varela's conviction that the Buddhist tradition and Western cognitive science have much to gain from each other provided an existential and spiritual dimension to his work, which became the subject of *The Embodied Mind: Cognitive Science and Human Experience* (1991). As previously noted, during the 1970s and 1980s, Varela served on the faculty of the Naropa Institute. During this period, he was a Fellow of the Lindisfarne Association. As a scholar in residence at Lindisfarne, supported by social philosopher William Irwin Thompson, Varela contributed to and learned from this intellectual, contemplative community. Their mission was to promote exchange in the humanities and sciences and spiritual and planetary transformation through interdisciplinary practices such as Yoga, Buddhism, Sufism, esoteric Judaism, and Christianity.

Per Thompson (2001b), through his involvement with the Lindisfarne community, Varela associated with luminaries such as systems theorist-epistemologist Gregory Bateson, who taught at the Humanistic Psychology Institute (now Saybrook University); James Lovelock, father of the Gaia theory; Michael Murphy, founder of Esalen Institute; musician Paul Winter; the mystic David Spangler; Zen Buddhist roshi-anthropologist Joan Halifax, founder of the Ojai Foundation and collaborator with Stanislav Grof; and Joseph Campbell and Alan Lomax who pursued LSD research. Among William Irwin Thompson's influences were British philosopher Alfred North Whitehead, Swiss cultural historian Jean Gebser, the mystic Rudolf Steiner, Vedic philosopher Sri Aurobindo Ghose, and Hindu philosopher Paramahansa Yogananda.

In Varela's "Reflections on the Chilean Civil War" (1979b), written for the Lindisfarne Association's *Newsletter 8*, he depicts his existential crisis and epistemological views of science:

> Chile was a process of understanding in the midst of traumatic social transformation… which had acquired a mythical connotation…. The Civil war gave me the experience that epistemologies are not something abstract to be given over only to historians of science; epistemology creates the kind of world that we live in and the kind of human values that we have. (p. 15)

His personal reflection continues,

> …. by 1973, they [the Chilean government and popular front] could not agree on anything, the time of day or the color of the sky… the polarity created a continual exaggeration of the sense of boundary and territoriality… which was the time at which things began to get very confusing. (p. 16)

Varela describes his sense of doom without understanding what any of this was about:

> … absolutely and completely chaotic…. I began to see tanks rolling down the streets, wagons loaded with soldier… war planes flying over the city… bullets screaming over your head. There is no sense in which we knew what is happening… there are no instructions, no government… there is simply no hope… and a strange sense that you don't know when your last moment will be. (pp. 17–18)

Varela describes the insight he derived from this experience as the "Logic of Paradise," a completely inverse understanding of the connection between world view, political action, and personal transformation:

> For me, the ground had been pulled from under me. Nothing was left to hold on to. At the same time a funny and contrary process happened; as things got more and more chaotic, the evidence of what a war is. There was a strange form of clarity coming more and more, a strange form of understanding, which I can't really express. I suppose it is somewhat like a semi-dream state. At the same time it was very real. (p. 18)

Varela reframed his existential crisis through an understanding of Buddhist epistemology, which enabled him to actualize and embody the polarity and relativity of human existence:

> … unless I was unable to cut through my sense of identity and attachment and identification with what I believe are my ideas, my things, my territory, my limits, I had no hope of understanding…. unless you build on the foundation of working with that sense of spirituality, there is simply no hope of understanding. I have found, for myself, expression of that understanding in Buddhist practice. I cannot separate that practice, that sense of working with the contemplation of how my mind and my actions generate and operate…. This is why I become so passionate about epistemology…. It is not an abstract proposition… when I say that we must incorporate in the enactment, in the projecting out of our world views, at

> the same time, the sense in which that projection is only one perspective, that it is a relative frame…. My conviction is that we must try to see to what extent our political views and our projections on the world can express this form of relativity, the fact that every position we take will also contain the opposite one. (p. 19)

To briefly continue with Varela's biography: he spent a year, 1979–1980, in New York at the Brain Research Laboratories of the New York University School of Medicine, and returned to Chile from 1980 to 1985 (with a year, 1984, as a Visiting Senior Researcher at the Max Planck Institute for Brain Research in Frankfurt). In 1986, he moved to Paris and affiliated with the Institute des Neurosciences and CREA. It was at this time that he met his wife Amy Cohen and began working on *Embodied Mind* with Thompson and Rosch. In 1988, he was appointed Director of Research at CNRS, a position he held until his death in 2001 (Thompson).

Varela pursued two main complementary lines of work: experimental studies using multiple electron recordings and mathematical analysis of large-scale neuronal integration, during cognitive processes, and philosophical and empirical studies of the "neurophenomenology" of human consciousness (Varela, 1996). With colleagues in Paris, he showed that the human perception of meaningful complex forms (high-contrast faces or "Mooney figures") is accompanied by phase-locked synchronous oscillations in distinct brain regions (Rodriguez et al., 1998). He also presented a viewpoint he called the "brainweb," that the emergence of a unified cognitive moment depends on large-scale integration (Varela et al., 2001). Additionally, he published technical, experimental, and mathematical papers on nonlinear dynamical analysis of brain activity including studies on the prediction of seizures in epileptic patients prior to the onset of symptom (Martinerie et al., 1998; Schiff, 1998) (Thompson).

Varela clearly believed that scientific research needs to be complemented by detailed phenomenological investigations of human experience as it is lived and verbally articulated in the first person. He published phenomenological studies of human consciousness (Varela, 1999; Varela & Depraz, 2000), including a meditation on his illness and the phenomenology of the experience of organ transplant (Varela, 2001). He also coedited *Naturalizing Phenomenology* (Petitot et al., 1999) and *The View from Within* (Varela & Shear, 1999).

To his credit, Varela was a founding member of the Association for the Scientific Study of Consciousness (ASSC) and a supporter of the Center for Consciousness Studies at the University of Arizona in Tucson and served on the editorial board of the *Journal of Consciousness Studies*. He was instrumental in creating the journal *Phenomenology and the Cognitive Sciences* and was to serve as its Consulting Editor at the time of his passing (Thompson).

Varela believed that the study of consciousness had potential for inciting a major revolution in what science is about and how it is practiced by going beyond subject-object dualism to the embodied, enactive self. Varela's Mind and Life Seminars, his exploration of altered and intuitive states of consciousness with his holiness the Dalai Lama (Varela, Ed., 1997), and his research project in neurophenomenology opened an interdisciplinary dialogue suggesting the future direction of a more person-centered science, and the subject of this volume.

References

Garcia, Varela, F. J. (1970). *Insect Retinas: Visual processing in the compound eye*. (Unpublished doctoral dissertation). Harvard University, Cambridge, MA (HU 90.9880.15).

Gendlin, E. T. (1978a). *Focusing*. New York, NY: Bantam Books.

Gordon, S. (2009). Radical empiricism and the neurophenomenology of time. Unpublished working paper.

Heidegger, M. (1927/1962). *Sein und zeit*. (J. Macquarrie and E. Robinson, Trans.). [Being and time]. New York, NY: Harper and Row.

Husserl, E. (1950/1966). *Phänomenologie des inneren zeitbewubtseins*. (J. S. Churchill, Trans.). [The Phenomenology of Internal Time Consciousness]. Bloomington, IN: University of Indiana Press. (Original work published 1893–1917)

Husserl, E. (1962). *Ideas: General introduction to pure phenomenology* (W. R. B. Gibson, Trans.). New York, NY: Collier Books. (Original work published 1913)

Husserl, E. (1977). *Phenomenological psychology* (J. Scanlon, Trans.) The Hague, Netherlands: Martinus Nijhoff. (Original work published 1925)

Husserl, E. (1991). *Collected Works. Vol. 4: On the phenomenology of the consciousness of internal time* (1893–1917) (J. Brough, Trans.). Dordrecht: Kluwer Academic.

James, W. (1890). *The principles of psychology* (2 Vols.). New York, NY: Henry Holt.

James, W. (1912). *Essays in radical empiricism*. New York, NY: Longmans, Green and Co.

Martinerie, J., Adams, C., Le Van Quyen, M., Baulac, M., Clémenceau, S., & Renault, B. (1998). Epileptic seizures can be anticipated by non-linear analysis, *Natural Medicine, 4*, 1173–1176.

Maturana, H. R., & Varela, F. J. (1973), De máquinas y seres vivos: Una teoría de la organizacíon biológica (Santiago: Editorial Univeristaria).

Maturana, H. R., Varela, F. J., & Uribe, R. (1974). Autopoiesis: The organization of living systems, its characterization and a model. *Biosystems, 5*, 187–196.

Maturana, H. R., & Varela, F. J. (1980). *Autopoiesis and cognition: The realization of the Living. Boston Studies in the Philosophy of Science, 42*. Dordrecht, Holland: D. Reidel.

Merleau-Ponty, M. (1962). *Phenomenology of perception*. New York, NY: Routledge & Kegan Paul. (Original work published 1945)

Merleau-Ponty, M. (1964). *Le Visible et l'Invisible*. Paris: Galland.

Petitmengin, C. (Ed.) (2009). *Ten years of viewing from within: The legacy of Francisco Varela*. Thorverton, UK: Imprint Academic.

Petitot, J., Varela, F. J., Pachoud, B., & Roy, J-M. (Ed.). (1999), *Naturalizing phenomenology: Issues in contemporary phenomenology and cognitive science*. Stanford, CA: Stanford University Press.

Robbins, B. D., & Gordon, S. (in press). Humanistic neuropsychology: The implications of neurophenomenology for psychology. In K. J. Schneider & J. F. Pierson (Eds.), *The handbook of humanistic psychology: Leading edges in theory, research, and practice* (2nd ed.). Thousand Oaks, CA: Sage Publications.

Rodriguez, E., Lachaux, G. N., Martinerie, J., Renault, B., & Varela, F. J. (1999). Perception's shadow: long-distance synchronization in the human brain. *Nature 397*, 340–343.

Schiff, S. J. (1998). Forecasting brain storms. *Nature Medicine, 4* (10), 1117–1118.

Taylor, E. (2009). *The mystery of personality: A history of psychodynamic theories*. New York, NY: Springer.

Thompson, E. (2001b). Obituary of Francisco J. Varela (1946–2001). *Journal of Consciousness Studies, 8*, 66–69). See also (http://psyche.cs.monash.edu.au/v7/psyche-7-12-thompson.html).

Thompson, E. (2007). *Mind in life: Biology, phenomenology, and the sciences of mind*. Cambridge, MA: Harvard University Press.

Varela, F. J. (1979a). *The principles of biological autonomy*. New York, NY: Oxford and Elsevier, North Holland Inc.

Varela, F. J. (1979b). Reflections on the Chilean civil war. *Lindisfarne Letter, 8*(winter), 13–19.

Varela, F. J. (1992/1999). *Ethnical know-how: Action, wisdom and cognition*. Stanford, CA: Stanford University Press.
Varela, F. J. (1996). Neurophenomenology: A methodological remedy for the hard problem. *Journal of Consciousness Studies, 3*(4), 330–349.
Varela, F. J. (Ed.). (1997). *Sleeping, dreaming, and dying: An exploration of consciousness with the Dalai Lama*. Boston, MA: Wisdom Publications.
Varela, F. J. (1999). The specious present: A neurophenomenology of time consciousness. In J. Petitot, F. J. Varela., B. Pachoud, & J-M. Roy (Eds.), *Naturalizing phenomenology: Issues in contemporary phenomenology and cognitive science* (pp. 266–314). Stanford, CA: Stanford University Press.
Varela, F. J. (2001). Intimate distances: Fragments for a phenomenology of organ Transplantation. *Journal of Consciousness Studies, 8* (5–7), 259–71.
Varela, F. J., & Depraz, N. (2000). At the source of time: Valence and the constitutional dynamics of affect. *Journal de lettre et de sciences humain 4*(1–2).
Varela, F. J., Lachaux, J-P., Rodríguez, E., & Martinerie, J. (2001). The brainweb: Phase synchronization and large-scale integration. *Nature Reviews Neuroscience, 2*, 229–39.
Varela, F. J., & Maturana, H. (1973). Mechanism and biological explanation. *Philosophy of Science, 39*, 378–382.
Varela, F. J., & Shear, J. (Eds.). (1999). *The view from within: First-person approaches to the study of consciousness*. Thorverton, UK: Imprint Academic.
Varela, F. J., Thompson, E., & Rosch, E. (1991). *The embodied mind: Cognitive science and human experience*. Cambridge, MA: The MIT Press.
Vaz, N. M., & Varela, F. J. (1978). Self and non-sense: An organism-centered approach to immunology. *Medical Hypothesis, 4*(3), 231–261; 263–267.

Note

1. Gestalt psychology, which defines personality as always greater than the sum of its parts, originated as a laboratory-based theory of perception, later becoming a science of the totality of experience. Max Wertheimer, Kurt Koffka, Wolfgang Kohler, and Kurt Lewin were prominent influences. Based on the phenomenological perspective, this movement began in Germany before World War I and extended beyond perception to the problems of learning, cognition, motivation, and personality. One of the more important circles in New York that had embraced dynamic psychology flourished around the anthropologists at Columbia, emanating from Franz Boas, Ruth Benedict, Margaret Mead, and by association Gregory Bateson (Taylor, 2009).

Enactive Cognition and the Neurophenomenology of Emotion

Brent Dean Robbins

> *... a mood assails us. It comes neither from the 'outside' nor from the 'inside,' but arises out of Being-in-the-world, as a way of such Being.*
>
> (Heidegger, 1962, p. 176)

If we consider the history of the theories of emotion and research in psychology, most accounts of emotion have been antithetical to the basic principles of phenomenological psychology. A conception of emotion that would be consonant with a phenomenological psychology would necessarily include certain essential qualities, in particular a nonreductive and holistic perspective, and by implication, a perspective that can account for the experience of emotion without merely explaining it away (Robbins, 2003; Robbins & Parlavecchio, 2006). Dominant approaches to psychology, in contrast, have tended to be mechanistic and reductionistic (Clegg, 2009). A genuinely phenomenological approach would need to account for emotion in a way that preserves the meaning of the experience as it appears within the life-world context of people living out their lives, rather than inferences about those lives drawn exclusively from ecologically invalid laboratory conditions, or experience-distant scaling procedures without strong associations to the lived meanings of affective phenomena. But while such an account should be nonreductive, to be truly holistic, it would have to avoid falling prey to inappropriate dualisms that split body and mind, subject and object. Does such an account of emotion exist today, and if so where? I would like to suggest that we are seeing the emergence of just such a perspective on emotion in perhaps the most unlikely of places: within the

B.D. Robbins, Ph.D. (✉)
Department of Humanities and Human Sciences, Point Park University,
201 Wood Street, Pittsburgh, PA 15222, USA
e-mail: brobbins@pointpark.edu

S. Gordon (Ed.), *Neurophenomenology and Its Applications to Psychology*,
DOI 10.1007/978-1-4614-7239-1_1,

field of cognitive neuroscience. Indeed, within this field, and the affective sciences in general, one can see a quiet revolution taking place: an experiential revolution in the research of emotion. I will argue that an enactive approach to cognition and a neurophenomenological approach to research is the most promising avenue for bringing this revolution to its complete fruition.

A perusal of the research and literature on the cognitive neuroscience of emotion bears this out: there is a sudden interest in the phenomenology of emotion, and this is because some cognitive neuroscientists are finally beginning to get the picture; one cannot explain the consciousness of emotion as a content of experience without necessarily presupposing the ontological primacy of the experience being explained (e.g., Gallagher, 1997; Ellis, 2006; Uhlhaas & Mishara, 2007). Let me put this another way: what neuroscientists are charged with doing is taking the experience of emotion and then explaining it in terms of cognitive processes and physiological events in the nervous system. These efforts from the beginning have always had as their goal to reduce the meaning of conscious experience to biomechanical explanations. But this approach has proved fatal for the project of a reductive cognitive neuroscience, because as soon as you explain something, you are presupposing the existence of whatever it is you set out to explain. If I say, "emotion is triggered, in part, by neuromechanisms in the amygdala of the brain's limbic system," this means that the amygdala's activity explains how emotion gets triggered, but what the explanation presupposes is an experience of emotion that is in need of explanation. But what is this experience of emotion? Clearly, from the first-person perspective of the person undergoing the emotional experience, this is not remotely experienced as an event in the brain. As the phenomenologist understands it, the first-person experience of emotion is the transformation of a world (Sartre, 2003). Neuroscientists who ascribe to phenomenology are beginning to come to terms with the fact that their neuromechanisms are tied to a world of experience that cannot be accessed with their instruments, but require careful experiential description.

Series of Revolutions

To put this discussion into an historical context, let us think about the sequence of revolutions in psychology over the past century. The cognitive revolution can be traced back to the 1950s and characterized as a counterrevolution against behaviorism (Baars, 1986; Gardner, 1987), which, among experimental psychologists in the USA, had become something like a language police, blacklisting young psychologists who attempted to use anything resembling what Skinner and colleagues referred to as "mentalistic language." The science of psychology was restricted to behavior and behavior only. The cognitive revolution probably originated with Jerome Bruner (1961; Bruner, Goodnow, & Austin, 1956), whose work centered on describing, understanding, and explaining human meaning. Bruner's revolutionary research, coined the "New Look," found various evidence to support the fact that perception is not merely a passive process but rather an active process of interpretation (Wertz, 1983). It was at this same point in history that humanistic psychology began its own revolution in psychology, which was consonant with Bruner's search for meaning (Wertz, 1994). But then came Chomsky, Minsky, and others, and the cognitive movement became more mechanistic and reductive and less concerned with human meaning, instead reducing meaning to "information processing," but it was a step forward (Bruner, 1993).

The second revolution in psychology has been coined the "affective revolution." In *The Handbook of Affective Sciences*, Joseph P. Forgas (2002) wrote: "The past two decades saw something like an "affective revolution" in psychological research. Indeed, one could argue that most of what is known about the influence of affect on attitudes and judgments has been discovered since the early 1980s" (p. 596). Indeed, if we look to the literature on emotion, the most substantial and lasting scholarship in this area rarely hearkens back further than the 1980s. Up until that time, cognitive science had primarily been informed by a computer metaphor of the mind, and of course computers do not have feelings, so there was not much attention paid to the affective dimension of the mind.

The first great rumblings of an affective revolution can probably be found in the work of Berkeley philosopher Hubert Dreyfus (1972), a phenomenologist who boldly put Heidegger and Merleau-Ponty in dialogue with analytic philosophy and cognitive science. Proponents of artificial intelligence (AI) were literally enraged by the publication of Dreyfus' hard-hitting critique of the AI project, *What Computers Can't Do: A Critique of Artificial Reason.* What Dreyfus demonstrated is that all work in AI research, in their attempts to simulate human cognition, was predicated on the basic assumption that humans, in some fundamental way, process information in ways that computers can emulate. He showed that this was no small assumption, because all computer-based information is explicit, discrete, linear, rule based, and definitive, while human thought does not seem to operate in this way. They assumed human beings operated like general-purpose symbol-manipulating machines, and this requires a variety of assumptions: that at some level people operate in a digital manner, that all thought is calculation, that all knowledge can be formalized, and that the world consists of context-free facts. These biological, psychological, epistemological, and ontological assumptions were made explicit by Dreyfus and criticized without mercy. Based on his grounding in phenomenology, Dreyfus was able to wonderfully articulate an alternative set of assumptions in which the body is understood to be the basis for intelligent behavior, in which orderly behavior must always be understood within the context of a situation, and in which the context of interpretation of the world must be situated within a uniquely human context that includes uniquely human needs. It was a brilliant argument, which still holds up well today, and anyone paying attention could see that the affective revolution was already there, implicit in his argument, as was the newly emerging experiential revolution. The book also contains what is implicitly a humanistic answer to the reductive scientism of AI and cognitive neuroscience that does not throw out the baby with the proverbial bathwater, which is what makes it brilliant.

The affective revolution came into full effect by the 1990s, when Richard Lazarus' pioneering work on stress and coping was developed into a full-fledged appraisal theory of emotion. In his seminal work *Emotion and Adaptation*, Lazarus (1994) explicitly referenced Heidegger as the inspiration behind his "relational" model of emotion, which he thought was a contribution beyond previous appraisal models of emotion and stress, in that it placed an emphasis on the emotion as a product of the relationship between the person and the world that results in meaning. For example, Lazarus elaborated:

> A central problem in understanding the role of cognitive activity in emotion is how meaning is achieved and changed in an adaptational encounter...I distinguished between two modes of appraisal: one automatic, unreflective, and unconscious or preconscious; the other deliberate and conscious. There may well be other modes...but the contrast between these two is especially important. The cognitive activity of observing and evaluating at both these levels may also be contrast with what Heidegger indentified as "being in a situation," and what Merleau-Ponty referred to as "embodied intelligence," both of which seem to be less distanced than when we act as an observer of ourselves. (p. 128)

Later in the book, Lazarus elaborated, "the concept of appraisal does not imply rationality, deliberateness, or consciousness of the cognitive processes involved in emotion" (p. 152). Drawing upon Heidegger's existential-phenomenological philosophy, Dreyfus argued that a person is not predefined, but discovers meaning over the course of living life, and that this meaning is primarily a nonreflective type of understanding. The possibility for this unreflective understanding is rooted in the human being's direct access to a meaningful world. Thus, for Lazarus, appraisal was not to be understood as an explicit, calculative type of cognition but as the existential meanings of situations, which are revealed ontologically through the structure of moods. For this reason, Lazarus stated unequivocably that he rejected a mechanistic approach to emotion and emphasized, on the contrary, a contextualist approach. These developments were very much in line with Hubert Dreyfus' critique of cognitive science and its applications in AI research.

The affective revolution, however, is not the end of the story: the experiential revolution is already implicit in the affective revolution. Both Dreyfus and Lazarus were inspired by Martin Heidegger and the phenomenological tradition, which informed their understanding of emotion as primarily an experience undergone by a person in relation to a world. But first, let us take a closer look at the state of affective science today. What we can glean from the literature is that there seems to be a major tectonic shift taking place in the area of cognitive neuroscience and in the passing away of an old paradigm and the emergence of a new one; the *experience* of emotion holds a special place—a place not amenable to biomechanical reduction.

The New Affective Science

In the early stages of affective neuroscience (e.g., in the era from the 1980s to the 1990s), most of the major figures in the field were openly hostile to inquiries into the structure of emotional experience. For example, Joseph LeDoux (1998, as cited in Lambie & Marcel, 2002) wrote, "The conscious feelings that we know and love (or hate) our emotions by are red herrings, detours, in the scientific study of emotions" (p. 18). Therefore, he went on to say, "Emotion researchers need to figure out how to escape from the shackles of subjectivity, if emotion research is to thrive" (LeDoux, 2000, p. 156, as cited in Barrett, Mesquita, Ochsner, & Gross, 2007, p. 374). But LeDoux and many of his colleagues are part of the old paradigm, which assumed that the study of subjectivity is antithetical to science. On the contrary, as I will argue, subjectivity is always already presupposed in any and all science and therefore must be accounted for on its own terms.

The old paradigm of affective neuroscience is built upon a dualistic understanding of cognition and emotion, which is one of the vestiges of the Cartesian mind-body split. In the field of neuroscience, this dualistic understanding manifests itself in the tendency to situate cognitive and affective functions in completely separate regions of the brain and nervous system, in which the systems are basically in opposition to one another. The earliest example is the model outlined by James Papez (1937), which has come to be known as the Papez circuit. Papez assumed that incoming sensory information was funneled by the thalamus into two separate streams—one stream being cognitive in nature and the other affective. The cognitive system was identified with the cingulate region of the brain, and the emotion center was believed to be situated in the mammillary bodies. Not long after, Paul MacLean (1949) modified the model of Papez to accommodate new knowledge about the role of the temporal lobe in emotion. He developed a triune theory of the brain, which identified more primitive areas of the brain as the seat of emotion. The striatal complex and basal ganglia were thought to be derived from the reptilian brain, and the limbic system was linked to emotional processes in more advanced mammalian species. The neocortex, which was identified as the seat of cognition, was thought to operate in such a way that it could, to some extent, regulate and control the more primitive structures of the brain.

LeDoux's (1996) model of the brain mirrors and reinforces the old paradigm. As with Papez and MacLean, emotion is identified with more primitive, subcortical regions of the brain. Positive emotion is linked to the nucleus accumbens, and negative emotion is thought to be generated by the amygdala, each of which receives copious input about incoming sensory information from the thalamic nuclei and sensory cortex before they project information back through the brainstem. Cognition, in contrast, is considered to operate once affect has already done its work—so here we have a conception of cognition as a Johnny-come-lately process, which, at best, functions to inhibit affective responses via pathways to the subcortical regions as a result of activity in the prefrontal cortex.

It was assumed that MacLean was correct until LeDoux built on this model, which previously suppressed the study of experiential constructs such as emotion.

The problem with the old paradigm is that the cumulative evidence does not add up. Research by Anderson and Phelps (2002) has shown that patients with damage to the amygdala are still capable of experiencing emotion, and this evidence directly contradicts LeDoux's model, which identifies the amygdala as the seat of negative emotion. What, then, is the role of the amygdala in emotion? New research suggests that "amygdala activation does not itself produce affective experience";

rather, it appears to "set the neural preconditions… for negative feelings to arise by influencing how sensory information from evocative stimuli is processed in the brain" (Barrett, Bliss-Moreau, Duncan, Rausch, & Wright, 2007, p. 73). In particular, the amygdala seems to increase sensitivity to negative stimuli, which in turn reinforces and enhances current and subsequent negative affectivity. In short, this line of research shows how structures of the brain previously thought to be exclusively linked to emotion are closely integrated with sensory systems and serve in the process of integrating sensation into perception, especially in terms of linking sensory objects to personal values—or to put it another way, emotion appears to operate in such a way that it integrates in perception, as one unified field, the spatial and temporal dimensions of a person's experience of other people and things, along with their felt value to the one who is doing the perceiving (Barrett, Bliss-Moreau et al., 2007).

Indeed, affect serves what were traditionally believed to be purely cognitive functions. These functions include the consolidation of memory, anticipation of the future, evaluation of the other people and things in the world, and, especially, through the integration and modulation of the process of the transduction of sensation into complex perceptions, the production of a personally meaningful world—a world articulated explicitly by phenomenology but which is lived primarily at a preverbal, implicit level of awareness. (Note: A more in-depth examination of the meaning of this preverbal, implicit level of awareness is developed below in the section on Heidegger's theory of moods).

By implication, then, cognition, emotion and perception are a well-integrated process that cannot be easily teased apart without making artificial distinctions that map onto neither brain function nor perceptual experience. Therefore, the thick tapestry of emotional experience can only be adequately linked to the function of the brain by careful, richly contextualized, and nuanced descriptions of emotional experience, articulated from a first-person perspective. This description must be careful to bracket or set aside assumptions about the nature of cognition and affect and their relationship to sensation, perception, and action.

The state of affective neuroscience today, therefore, suggests that neither the study of neuroanatomy nor the construction of hypothesized cognitive processes, which may or may not operate behind the scenes of experience, is adequate for a complete and satisfactory science of emotion. The reductive, mechanistic approach of contemporary cognitive neuroscience is unable to account for the very construct that is central to the everyday understanding of emotion as an *experience*. For example, we know from rigorous studies that physiological processes are only modestly associated with the experience of emotion (Mauss, Levenson, McCarter, Wilhelm, & Gross, 2005). Therefore, something more is needed for a science of emotion: a rigorous approach to the description of the experience of emotion, as it is felt from a first-person perspective. That is why, today, we are seeing the fourth revolution in the history of contemporary psychology. The behavioral revolution gave way to the cognitive revolution, and then the cognitive revolution produced the affective revolution. Today, the affective revolution is necessitating the next logical step: a revolution in the study of emotional experience.

The Next Step

Within the new experiential revolution, there seems to be two broad pathways: one pathway is that of phenomenology and the other pathway is neurophilosophy. The phenomenological pathway encompasses both humanistic and human science approaches, and, arguably, it is the only tenable approach to the study of experience. The latter approach, neurophilosophy, as exemplified by the work of Patricia Churchland (1986), is another version of reductive scientism. In my opinion, the path of Churchland's neurophilosophy, if taken, would be an unnecessary detour that would set the field back several decades before it finally settled on the only genuine method for the study of experience: the phenomenological method.

Shaun Gallagher (2007) has identified the basic tenets of Churchland's position on neurophilosophy. Her tactic, and one that seems to be consistent with much of current research in the neurosciences, is to start from common sense and folk wisdom and then to test these ideas experimentally. Eventually, in the end, the idea would be to replace folk psychology and introspective phenomenology with reductive, neuroscientific explanations. Her idea, then, is to exchange the language of phenomenal experience for the mechanistic language of biomechanics.

The problem with Churchland's approach is that it is incapable, based on its method, to explain how nonconscious mechanisms can become consciously experienced phenomena—her approach takes this fact for granted as its starting point and so presupposes what it is supposed to explain. Neuroscience and neurophilosophy has set itself the task of explaining phenomenal experience—that is, as Gallagher (2007) explains, "not only how it is generated but what it is like" (p. 302). Yet what remains in need of explanation are the specific features of consciousness, or phenomenal experience, that these representations or neural activations are supposed to be producing. Neurophilosophy attempts to explain the processes of the brain by an appeal to folk wisdom—in other words, it tests common sense understandings of categories of experience and attempts to correlate these categories with mechanisms in the brain. To answer these types of questions, however, neuroscience and folk wisdom are woefully inadequate. To address these questions, at the very least, the field is in need of a descriptive phenomenological method—some kind of systematic, rigorous, and controlled description of lived experience. These controlled descriptions would, in turn, provide constraints upon neurological explanations and clarify what is in need of explanation. And they would do so in a way that would not, and could not, without serious contradiction and coherency, lead down the path of reductionism. As noted above, reductionism is a problem because a reductive approach to explaining experience presupposes what it is supposed to explain. The study of consciousness is itself dependent upon the consciousness of the philosopher or scientists if it is to even get started. Therefore, a coherent approach to the experience of emotion would have to be a holistic and integrated perspective that appreciates what is unique about human experience and behavior without forgetting that it is fundamentally rooted and always already situated in an embodied and worldly existence.

Phenomenology therefore has another advantage over traditional reductive methods. In contrast to reductive approaches, phenomenology need not rely upon folk wisdom to generate hypotheses, because "phenomenology, in its practice of phenomenological reduction, suspends the common senses attitude ("natural attitude") of folk psychology and aims to develop precise descriptions of experience as it is" (Gallagher, 2007, pp. 130–131). The word "reduction" here should not be confused with "reductionism." Whereas biological reductionism is aimed at eliminating the problem of consciousness by explaining it as composed of a simpler, physiological, and mechanical event in the nervous system, the phenomenological reduction is an attitude that sets aside assumptions about the reality of the natural world in order to observe precisely how experience unfolds before us. Such a radical step, rooted in the philosophy of Edmund Husserl, is able to step back and observe experience with less reliance upon traditional metaphysical categories such as subjectivity, objectivity, mind, and body, which are presupposed by most cognitive science. The step of the phenomenological reduction is absolutely necessary because folk psychology is riddled with dualistic presuppositions, including the mind-body split and the cognitive-affective divide, which has plagued cognitive science and neuropsychology for more than a century. Only a radical phenomenological reduction, which brackets such categorical distinctions, has a chance at truly building an experientially congruent description of affective phenomena.

At the same time that phenomenological description can bracket the natural attitude and place it in abeyance, the results of phenomenological research can be placed back into dialogue with correlational and experimental research on cognition, affect, and the brain. This kind of dialogue between phenomenological and experimental research is described by Christopher Mruk (2008) as an "integrative phenomenological" approach. When restricted to a dialogue between cognitive neuroscience and phenomenology, Francisco Varela, Shaun Gallagher, and others have referred to it as neurophenomenology (e.g., Gallagher, 2007; Varela, 1996; Varela, Thompson, & Rosch, 1991).

In addition, there has already been some success in the development of a mixed-method, phenomenological approach to understanding experiences of emotion. For example, in my phenomenological research on embarrassment, I studied the first-person narratives of participants, which yielded the eidetic finding that, in essence, what it means to be embarrassed is to find oneself subject to unwanted exposure (Robbins & Parlavecchio, 2006). The finding is "eidetic" in that it utilized the second step of the phenomenological method following Husserl's phenomenological reduction. This second step, called the *eidetic reduction*, varies aspects of the phenomenon until what is left are only those aspects of the phenomenon that are essential to its meaning. What remained as essential to embarrassment, after careful qualitative analysis, is the essential meaning of embarrassment as "finding oneself subject to unwanted exposure." Then, in a subsequent study, I used a quantitative, experimental methodology to vary this theme in second-person perspective narratives (Robbins & Fleming, 2008). Using first-person descriptions of embarrassment by participants from a prior study, the pronouns in the narratives were changed from first-person to second-person pronouns. This had the effect of helping participants read the descriptions in a way that allowed them to identify with the experience being described. After reading the narratives, participants numerically rated how embarrassed they would be if they actually experienced the situation described in the story. When participants read a narrative that included the theme of unwanted exposure, they, on average, rated it to be significantly more embarrassing than when the theme of unwanted exposure was removed. The results clearly supported and validated the previous phenomenological findings.

Likewise, in recent research published in *The Humanistic Psychologist*, I studied the phenomenology of suppressed laughter and again was able to validate these findings using a quantitative method that varied thematic elements of second-person narratives (Robbins & Vandree, 2009). The qualitative, eidetic analysis identified essential themes of suppressed laughter. A key finding was that suppressed laughter is a profoundly social phenomenon. A social context that included at least three actors was necessary for the suppressed laughter to occur. In addition, the social context necessitated a situation in which laughter was inappropriate, such as a classroom or funeral. A follow-up experimental study used second-person narrative descriptions of suppressed laughter and varied the themes of the social context. As predicted based on the eidetic analysis in the former study, the experimental results found that situations were rated as more humorous when laughter was inappropriate to the social context and also when a friend was present.

These mixed-method studies of embarrassment and laughter are a few examples of what could be a new paradigm of mixed-method phenomenological research. Invariant thematic descriptions of phenomena, by the use of this method, can be better generalized beyond the small data set required for careful, methodical phenomenological description. These findings, therefore, provide a fairly secure ground upon which to build a description of emotional experience that can be put into dialogue with research in neuroscience. For example, the studies of embarrassment predict that genuine experiences of embarrassment and laughter involve a basic capacity for the social cognition of other minds, as well as the imagined perception of one's self through the perspective of the other person. Neuroscience research shows that different brain regions are involved in the process of self-awareness from a first-person perspective versus third-person perspective awareness of other persons (Banissy, Kanai, Walsh, & Rees, 2012; David et al., 2006; Vogeley, May, Ritzl, Falkai, Zilles, & Fink, 2004), and the capacity for third-person perspectivity is a developmental milestone that emerges in conjunction with the expression of embarrassment (Moriguchi, Ohnishi, Mori, Matsuda, & Komaki, 2007). Consequently, emotions that involve complex social cognition, such as embarrassment and suppressed laughter, ought to involve brain regions closely associated with these social cognitive processes. For example, based on research on the development of the perception of other minds over the course of child and adolescent development, we might predict that embarrassment and suppressed laughter hinge upon activation of the superior temporal sulcus (STS), the temporal pole (TP), and the medial prefrontal cortex (MPFC) (Moriguchi et al., 2007). Future imaging studies may bear this out or, if not, would lead to further questions in order to refine our understanding of these complex human emotions.

Using a different approach that is more about linking phenomenal experience to cognitive and brain function, Lutz, Lachaux, Martinerie, and Varela (2002) combined phenomenological descriptions by participants with EEG recordings and behavioral measures and were able to show, as described by Gallagher (2007), "correlations between experience (attention level), reaction times, and dynamic descriptions of the transient patterns of local and long-distance synchrony occurring between oscillating neural populations, specified as dynamic neural signature (DNS)" (pp. 123–124). For more examples of this type of research, see Gallagher and Sorensen (2006).

So we are situated at a new frontier in affective neuroscience, and standing at this threshold, phenomenology has a unique opportunity to assert its relevance to the problems that are facing this highly influential subfield of psychological study. Yet some may question why it matters. Why should psychologists and neuroscientists be interested in a phenomenological approach? Within a reductive neuroscientific approach, the aim essentially is to eliminate psychological constructs from the picture. However, in contrast, neurophenomenology demonstrates why such a reductionism is neither possible nor desirable, because the structure of experience is more ontologically primary than physiological facts. To demonstrate this, I will turn to the philosophical work of Maurice Merleau-Ponty and Martin Heidegger. By gaining an understanding of the ontological primacy of experience, we can later return to address the benefits of an approach to emotion that gives primacy to emotional experience prior to examination of the physiological correlates of emotion.

Beyond Reductionism via Neurophenomenology

Of particular interest to this discussion is Maurice Merleau-Ponty's first major work, *The Structure of Behavior* (1983), the purpose of which is "to understand the relations of consciousness and nature: organic, psychological or even social. By nature we understand here a multiplicity of events external to each other and bound together by relations of causality" (1983, p. 3). When we start with nature understood in this way, it gives rise to what is sometimes called the "explanatory gap"—that gaping abyss between consciousness and the natural world. Merleau-Ponty identifies behavior as a third term between these classic dualities—although a better translation from the French is *comportment*. But behavior in Merleau-Ponty's work here is very different than what we usually think of as behavior—and through his understanding of behavior, Merleau-Ponty argues, we can arrive finally at an understanding that consciousness must not be seen as a psychological reality nor as a cause but, rather, as *structure*. Hence, the title of the book is *The Structure of Behavior*. When you hear the term structure here, think *form*. Merleau-Ponty writes:

> There is form whenever the properties of a system are modified by every change brought about in a single one of its parts and, on the contrary, are conserved when they all change while maintaining the same relationship among themselves. (p. 47)

Therefore,

> form possesses original properties with regard to those of the parts that can be detached from it. Each moment in it is determined by the grouping of the other moments, and their respective value depends on a state of total equilibrium the formula of which is an intrinsic character of 'form.' (p. 91)

Now, if behavior is a structure or form, then this means that mechanical thinking really cannot deal with it. Mechanical thinking cannot deal with form because it is a process of taking wholes and exhaustively analyzing them into parts that are externally related to one another. Yet this is precisely *not* what behavior is if behavior is structure or form. Because structure or form is the basis of both life processes (the province of biology) and mental processes (the province of psychology), mechanical thinking cannot understand such processes (Thompson, 2007). Why? First, because we cannot break down these structures into parts without losing the form and, therefore, without losing the very structure that is the behavior by definition. Secondly, mechanical analysis fails to appreciate or disclose the way in which behavior and biological forms are constituted by the reciprocal determination of whole and part. What is needed is a different kind of thinking than mechanical thinking—what we need is what Merleau-Ponty calls *dialectical thinking.* Dialectical thinking permits analysis of dialectical relations.

Here is how Evan Thompson (2007) defines dialectical relations:

> … (1) A determines B, and B determines A (bi-directional dependence or reciprocal determination); and (2) neither A nor B is analyzable into discrete, causally efficacious elements that stand in a one-to-one correspondence (nondecompsability). Furthermore, dialectical relations are dynamic, not static. Hence (3) A alters B, and B alters A; (4) A is altered by B as determinant of B, and B is altered by A as determinant of A; and (5) it makes sense derivatively to speak of A making what A is via B, and B making what B is via A. Given these kinds of close interdependencies, A and B can also be regarded as parts of a larger global whole or pattern when they are dialectically related. Hence (6) what A is a part of is what B is a part of. (pp. 68–69)

If a living, biological organism or a human being is understood in terms of dialectical relations, then, says Merleau-Ponty (1983), we come to realize that "The genesis of the whole by composition of the parts is fictitious. It arbitrarily breaks the chain of reciprocal determinations" (p. 50). What about physical stimuli? Are not organisms causally determined by physical stimuli pressing upon them from the outside and within? Merleau-Ponty rejects this conception. Rather, physical stimuli influence the organism not in a linear, cause-and-effect chain of events, but rather by eliciting from the organism a global response. They are "occasions," not "causes." In other words, "the reaction depends on their vital significance rather than on the material properties of the stimuli" (p. 161). In other words, we can say that the organism is responding to an informational stimulus, not a physical stimulus—an informational stimulus in the sense that the stimulus is informed by (the form or structure of) the organism. Within this context, a stimulus is a triggering condition, not an efficient cause.

The implication is that organisms do not exist in a linear chain of cause-and-effect relations that can be understood mechanically. They exist in a context of significance in relation to meaning. In other words, meaning is intrinsic to the belonging together of organism and milieu, and thus behavior cannot be reduced to relations of mutual exteriority. How about that for nondualism? There is not an organism over or against an environment, but a single structure within which the organism and milieu are expressed as behavior, or comportment. Therefore, organism and milieu are not related in a stimulus-reaction relationship but, rather, in terms of situation and response. Brain, body, and environment—all are necessarily integral, essential constituents, of behavior, and therefore it cannot be reduced to something located just inside the nervous system. There is no homunculus.

Once we have established the structure of behavior, Merleau-Ponty's next task is to articulate the emerging complexity of three ascending orders of the natural world: the physical, the vital, and the human.

The Physical Order. Even at the physical level, there is form or structure. Think of a soap bubble, or a similar physical structure, in which we have a structural stability. Even here, says Merleau-Ponty, analytical reductionism is not adequate to the task. Merleau-Ponty writes that such physical forms possess "internal unity inscribed in a segment of space" and resist "deformation from external influences by its circular causality" (p. 137)—thus, the physical form is "an individual." These insights have been borne out by contemporary science, such as Rene Thom's (1975) catastrophe theory and Jean Petitot's (1992) morphodynamical "physics of phenomenality" (Thompson, 2007).

The Vital Order. What about the vital order? Here we have the emergence of living structures in the physical order. The shift here is a shift from matter to life. How does a living structure differ from physical structures? They do so in three ways, according to Evan Thompson (2007):

1. A living cell dynamically produces and maintains itself through the continual chemical synthesis and breakdown (anabolism and catabolism) of material compounds, including those that make up its own membrane boundary. Further, they endogenously control and regulate their own external boundary conditions, whereas physical structures do not. Also, as living structures, they are self-producing and self-regulating unities (p. 73).
2. The material and energetic demands of this entire process orient the cell of necessity toward the environment, not simply in the sense of real and present conditions but also in the sense of conditions that need to be actualized (effect or procured)—in other words, virtual conditions. Organisms shape the physico-chemical environment into a milieu (an *Umwelt*) (p. 74).
3. Whereas physical structures can be expressed by a law, living structures have to be comprehended in relation to norms (p. 74). Or, as Merleau-Ponty (1983) writes, "each organism, in the presence of a given milieu, has its optimal conditions of activity and its proper manner of realizing equilibrium," and the modification of the milieu by the organism is done in accordance with the "internal norms of its activity" (pp. 148, 154).

The above qualities can be found even in simple bacterial cells, such as *E. coli*. They are qualities of a new order of nature that is qualitatively distinct from the physical order, because within this vital order, the organism is an individual whose relationship to the environment is meaningful and normative.

The Human Order. Now to the third order: the human order. The most typical structures and forms of behavior in the symbolic order are symbolic. Symbols and symbolic structures do not exist in isolation but belong to systems of symbols. In these systems, each symbol is related not simply to what it symbolizes—the thing or event for which it stands—but also to other symbols. Within this order, it becomes possible to express in a wider variety of ways and to represent the same thing in multiple ways. With these new abilities comes the possibility of "representing" as having an invariant structure across a diversity of aspects and perspectives. Only in the symbolic order does a yo-yo, for example, remain the "same" yo-yo despite multiple changes across background contexts. Only within this symbolic order can a doorknob remain a doorknob even for a double amputee, for example. In short, the symbolic, human order makes it possible to grasp something as an object—at least in the sense of phenomenology, by which we mean that which remains invariant throughout variations in perspectives and which is understood to be graspable by other people even if no longer by one's self. The milieu has changed. No longer merely the situation-response of the vital order, the human milieu is a matter of perceived situation-work. Here, reality is intersubjective, and things become "use objects" by which they are endowed with culturally constituted meanings.

A unique feature of the human milieu is its ambiguous status, which results from its dialectical relations. The social and cultural saturation of the human milieu is created by human beings themselves. We can commit ourselves to cultural perspectives as a result of symbolic forms—and even still, we nevertheless tend to refuse to allow ourselves to become completely identified with them. We are always trying to pass beyond these identities, always attempting to create new things, always attempting to be generative.

We have not yet directly addressed the issue of the explanatory gap. So now we turn to the issue of consciousness and the structure of behavior. There are two implications of the above analysis: the traditional notions of exteriority and interiority become highly problematic. The natural world, in a certain sense, can be seen to be not simply a matter of exteriority but has its own interiority, sharing structural analogs to the mind. And, likewise, what is usually considered to be pure interiority, the mind, is engaged with the world as form or structure, thus not merely interiority.

In other words, Merleau-Ponty's work on structure and comportment breaks from the Cartesian understanding of nature, which is to understand nature as "a multiplicity of events external to each other and bound together by relations of causality" (p. 3). Instead, for Merleau-Ponty, forms or structures in nature are analogous to a melody, in which all of the particular notes are what they are in their relation to all the other notes in the score. But, interiority does not yet exist at the physical level: it emerges at the organic or vital level, where life emerges. Interiority arises because an organism, by definition, is a being, which self-produces an inside—literally a body of some kind—which regulates a relationship with the environment in terms of some kind of norm or equilibrium.

Now, here is the interesting part: without an inside, there can be no outside. So, technically and speaking very precisely, prior to the emergence of life, there is no inside nor outside of anything—not really. There is no interior nor is there any exterior. These types of relationship only become possible at an organic, vital level of existence. Now, that is some serious nondualism. Inside and outside are co-emergent.

However, there is an important point to be made here. The co-emergence of inside and outside is not a symmetrical relationship (Thompson, 2007). The production of an inside takes certain precedence, because it is only by virtue of the creation of a body with an inside to regulate the surrounding milieu that an "outside" emerges. And this is why we cannot think of life in terms of a mechanism without losing something essential to what life is—maybe *the* most essential thing about life: that it has a kind of *inner* life. You might even say that it has a soul or psyche.

Now, given this scenario, how are we then to think of consciousness? What becomes of the Cartesian notion of mind? Well, unlike Descartes, we no longer think about consciousness as some kind of interior state of the mind or brain—at least not in the sense of something in a chain of linear cause-and-effect relations—a chain of sensory input and motor output, as we find in most introductory psychology textbooks. Rather, mind is understood as comportment, behavior, or form and structure—a perceptual and motor attunement to the world. And in the case of the human order, we go one step beyond this and include a particular kind of attunement to the world, in which the environment takes on the character of being symbolic and in which the actions of others are understood intuitively to be intentional. Thus, for example, emotions such as embarrassment cannot be understood adequately from a purely reductive physiological perspective, because it by definition eliminates from its purview the social, experiential, and symbolic context in which such an emotion has its meaning and ground.

Naturalism and the Phenomenological Attitude. When we understand the mind and nature in this nondualistic way, what happens to naturalism? And where does the phenomenological attitude come into play? In the phenomenological approach to philosophy, as articulated by Edmund Husserl (1982) and elaborated upon by Maurice Merleau-Ponty (2002), the first moment of the phenomenological process is what Husserl called the phenomenological reduction, or the *epoché*. In this special attitude, the phenomenologist suspends or brackets his or her assumptions about the natural world and its status. The phenomenologist places in abeyance all those assumptions we bring with us about what belongs on the subjective side of the equation of consciousness and what belongs on the objective side. The phenomenologist, in the attitude of the phenomenological reduction, simply describes how things show up in consciousness and their process of unfolding in experience over time. The second moment of the phenomenological method is the *eidetic* reduction. Here, the phenomenologist asks what constituents of a phenomenon are essential and what are nonessential by, for example, imaginatively varying these constituents in one's experience. By discarding nonessential constituents, the phenomenologist aims to arrive at invariant, thematic structures of transcendental consciousness, which Husserl thought were constitutive of all sciences and all human endeavors.

So why then is the phenomenological reduction necessary? Have we not thus far presented what seems to be a compelling argument from an attitude that comes from the empirical sciences—physics, chemistry, biology, and psychology? What need is there for phenomenology?

The phenomenological move becomes a necessary step, because Merleau-Ponty's (1983) argument reveals something quite startling, that we take a step back and examine the way in which form gets constituted as an object for cognition in the first place. In what manner does this form appear? What is the epistemological basis or ground for this appearance? Now, we are suddenly thrust into the necessity for transcendental phenomenology. We are now doing philosophy (not phenomenological psychology! not yet!) and must step back from the view of the empirical sciences to understand the meaning and ground for the empirical sciences.

From within the perspective of the phenomenological attitude, and the eidetic reduction, Merleau-Ponty (1983) draws the following conclusions (wonderfully summarized by Evan Thompson, 2007, p. 81):

1. The notion of form is borrowed from the perceived world.
2. The notion of form is encountered in physics only to the extent that physics refers us back to perceived things.
3. Hence, the notion of "physical form" cannot be the real foundation of the structure of behavior, in particular of behavior's perceptual structure.
4. The notion of form is conceivable only as an object of perception.

The conclusion is mind-blowing in its implications. To put it succinctly, naturalism as we know it requires the notion of form to even get started, yet form is irreducibly phenomenal. As a result, naturalism as we know it is incapable, in principle, of ever explaining matter, life, and mind, as long as it continues to purge subjectivity from its worldview and as long as it continues in its attempts to reconstitute subjectivity out of a subjectless nature. (Note: For serious skeptics, this argument can also be shown by way of quantum physics, e.g., Barr, 2006).

Without a careful consideration of what has just been done, it can appear as if Merleau-Ponty is making a case for metaphysical idealism. The argument could be mistaken for an argument that asserts a view that holds that physical forms are constructions out of a preexistent consciousness. And this is not the case. Actually, Merleau-Ponty is performing a transcendental line of thought. Any object has to be understood in its relation to the mental activity that intends it, but the object is not reducible to the mental activity that intends it. The physical world is not denied. What *is* denied, and undermined in a radical way, is the objectivist view of the physical world. In contrast to objectivism, the natural world is translatable into the sciences and into a world of predictable calculations, not because it is examining only the intrinsic properties of autonomous entities but because it holds true for any (human) active subject.

As Michel Bitbol (2003) elaborates:

> [Science] is not supposed to reveal anything of a preexistent underlying absolute reality, nor is it a more or less random aggregate of efficient recipes. Science is rather the stabilized byproduct of a dynamic reciprocal relation between reality as a whole and a special fraction of it. Defining this special fraction of reality qua subject is the reverse side of its actively extracting objectlike invariant clusters of phenomena. (pp. 336–337)

The upshot of this whole line of argument: for any object to be disclosed, consciousness is always already presupposed as that object's invariant condition of possibility. This results from the transcendental status of consciousness. In other words, we cannot step outside of experiencing subjectivity in a way that would permit us to map it onto an external reality purged of any and all subjectivity. Mind emerges from matter and life at an empirical level, but at a transcendental level every form or structure is necessarily also a form or structure disclosed by consciousness. With this reversal, one passes from the natural attitude of the scientist to the transcendental phenomenological attitude; one shifts into the *epoché*.

Where does phenomenological psychology fall into this big picture? Well, once we adopt the transcendental phenomenological attitude, then the experience of the lifeworld that is not within the purview of other empirical sciences is in need of descriptive elaboration, and empirical phenomenology becomes necessary to describe that lifeworld left out of the physical sciences. This phenomenal world or lifeworld also comes to be understood as having a certain ontological priority, since it is only through the lifeworld—the symbolic world and the vital world of inside-outside—that any science can get started. The lifeworld is that medium through which a transcendent world beyond experience can come to be known gradually through careful observation over long breadths of time in systematic observation of accumulating evidence—and nowhere else is this possible as far as we know, but in the human order.

For research on emotion, this revolutionary paradigm shift has very important implications. It suggests that careful and detailed description of the everyday lifeworld (not to be confused with folk wisdom and anecdotal explanations, which are abstractions from everyday engagement with the world) has to come first before we can press on to discuss physiology. Physiological events are themselves experienced through scientists engaged in everyday activities in the world, and so all perception of physiological events relies upon human receptivity. Integral to all human activity is mood or emotion, which enables the human agent to perceive how other people and things matter to him or her. And so emotion becomes the essential ground within which all cognition occurs so that the human being can decide what actions to take in order to cope with ongoing events in the world.

Enactive Cognition and the Phenomenology of Emotion

The "enactive approach" to cognitive science was first articulated in the book *The Embodied Mind: Cognitive Science and Human Experience* by Varela et al. (1991). This enactive approach, which is completely in line with the perspectives offered thus far, shares several key principles. The term *enaction* refers to the concept that any living human being is distinguishable from nonliving beings by virtue of its autonomy, agency, and ability to maintain its structure or form. Thus, living beings *enact* their own cognitive environment, or milieu. In addition, living beings with nervous systems enable the organism to maintain autonomy and form, in which the nervous system itself enacts its own meaningful patterns within an interacting system of neurons. With these kinds of nervous systems, living beings engage in cognitive activity that is integrally tied to full-bodied engagement with a perceptual world of activity. The coordination of perception and action in the nervous system is enacted as affective states and other cognitive activities in which environment and sensorimotor activity emerge as structurally and dynamically synchronized. Therefore, cognitive activity is not something merely going on "inside the head" (Noe, 2006, 2010), but rather is the emergent dynamic structure, or form, in which the living being comports itself within a milieu or environment it throws up before itself in its engagement with others and things. Finally, therefore, as we have already discussed, it is wrong to think of experience as epiphenomenal and equally problematic to conceptualize experience in terms of errant folk wisdom. Careful, phenomenological description demonstrates the ontological primacy of experience and the necessity of grounding physiological observation, including our understanding of the brain's physiology, within a phenomenological groundwork. As we have seen through the arguments of Merleau-Ponty and Thompson, experience has ontological primacy because it is only through experience that physiology comes into being as a body of thought within a human community. To give primacy to experience entails radically revising our categories of understanding for constructs such as cognition, emotion, and perception. One example of a radical, phenomenological investigation of human being is Heidegger's existential analytic in *Being and Time* (1962). We will see that it remarkably dovetails with recent advances in the cognitive science of emotion.

Heidegger's Theory of Moods

In *Being and Time*, Heidegger (1962) begins his analytic with an exploration of the existential condition of the human being as *being-in-the-world*. To understand what Heidegger means by world, he distinguishes between two modes of engagement with things, a "ready-to-hand" and a "present-at-hand." A present-at-hand mode of engagement is way of perceiving things that sees them as discrete objects externally related to one another and detached from ourselves. This is the manner of seeing of the scientific investigation of things. Because of the status of science in our culture, we believe that this way of seeing things is the only legitimate or ontologically valid way of seeing them. Heidegger reveals that, on the contrary, there is a more basic or primordial mode of engagement, which he calls "ready-to-hand," which is the way things appear when we are actively engaged with them in some kind of meaningful activity. In this mode of engagement, a tool such as hammer, for example, is felt to be an extension of the body, and the body self-effaces and conceals itself as the person becomes perceptually absorbed in the flow of the activity. This is the way we usually interact with things, and it is only when things break down, or when a tool is not working properly, that we are called to detach from our engagement with them and contemplate their properties in an uninvolved and reflective way.

When Heidegger uses the term "world," then, he is identifying it as an existential condition of possibility for human beings to be what they are. In the "world," we are typically engaged with things in a "ready-to-hand" way in which things appear within a field of "equipmentality." In other words, things make reference to other things within a larger network of activity so that we can talk about the "world" of a carpenter's workshop, or the "world" of a business executive, or the "world" of a medical doctor and immediately we can envision how such a world would appear if we were actively involved in it. Likewise, when studying moods and emotion, we can begin to talk about the "world" of anger, or the "world" of joy, and begin to get a sense of what it would be like to live in that field of significance, including the way other people, things, and our bodies would appear and feel in that context.

The concept in Heidegger's work most closely resembling what we call mood or emotion is the term *Befindlichkeit*, sometimes translated as "state-of-mind." As Gendlin (1978b) points out, this translation of *Befindlichkeit* into "state-of-mind" does not quite capture the original, German meaning Heidegger was going for. The German phrase *Sich befinden* translates roughly as "finding oneself" and has three allusions:

> The reflexivity of finding oneself; feelings; and being situated. All three are caught in the ordinary phrase, 'How are you?' That refers to how you feel but also to how things are going for you and what sort of situation you find yourself in. To answer the question you must find yourself, find how you already are. And when you do, you find yourself amidst the circumstances of your living. (Gendlin, 1978b, p. 44)

In the sense of the three above allusions, *Befindlichkeit* could be rather awkwardly translated into English as "how-are-you-ness," or even "self-finding."

Befindlichkeit differs in several ways from conventional thinking about mood and emotion. First, it does not refer to a merely subjective experience, since it is an ontological category and therefore prior to the epistemological abstraction that distinguishes subject from object. Secondly, *Befindlichkeit* is not merely a passive experience undergone by a person. Rather, it always already, whether explicitly or not, has an understanding of the world in the sense that one "finds one's self" situated in a context of significance in which one's projects are going well or badly, or even ambiguously, on their way to being fulfilled. Further, *Befindlichkeit*, as ontological, precedes the metaphysical distinction between cognition and emotion. Finally, *Befindlichkeit* belongs together with language, which is necessary for any explicit or implicit articulation. As Gendlin (1978b) writes, "We sense ourselves in situations with others, with an implicit understanding of what we are doing and with communication between us always already involved. A feeling is all that" (p. 45). Thus, as ontological, the existential of *Befindlichkeit* is prior to the metaphysical distinction between a self-contained self and other.

When moods are understood as an ontological category, they can be seen to reveal certain things about being-in-the-world. Namely, they reveal that we are thrown into the world without fully understanding why and where we are going to end up. They also disclose the world of our concern, and the field of equipmentality and its significance, as we discussed above, as well as being itself as a whole. And, finally, moods reveal "mattering," or how things and others matter and make a difference to us and our concerns. In addition, Heidegger's existential analytic of mood or *Befindlichkeit* has to be placed within the subsequent part of his analysis, in which mood is a coexistential with two other categories of human existence, understanding and fallenness. In having understanding, this means that human beings are always projecting possibilities for the self, and therefore, the "world" is primarily a world of possibilities yet to be actualized. Anything a person is engaged with in their current field of equipmentality, they are doing for the sake of the realization of some future possibility. So being-in-the-world is fundamentally temporal—it has the structure of time, in which human existence is oriented toward the future in the present moment, while the past recedes behind it. The past is represented by fallenness—our tendency to understand ourselves in terms given to us by our culture and by past experience.

From Heidegger to Cognitive Science

Remarkably, Heidegger's existential analytic, which reveals an ontological structure of the human being consisting of mood, understanding, and fallenness, has recently been mirrored by new insights in cognitive neuroscience. In an article titled "The Future of Psychology: Connecting Mind to Brain," Barrett (2009) explores the failure of reductive neuroscience to explain emotion. She attributes this failure to the imposition of folk wisdom onto human experience and the structure of the brain. Scientists studying the brain essentially buy into abstract concepts borrowed from folk wisdom, and this leads them down blind alleys. For instance, scientists have a tendency to take for granted the folk wisdom that cognition is categorically distinct from emotion, and therefore, they also assume that there must be distinct, anatomical structures to account for these distinct categories. Barrett rejects this assumption and engages in what is in essence a phenomenological reduction, even if she does not call it by that name. Whatever the case, she radically questions what it means to be in a mood and arrives at the tentative conclusion that all more complex psychological states, such as anger or joy, are built up from what she calls "psychological primitives" (Barrett).

The psychological primitives identified by Barrett include (1) valuation, salience, or affect; (2) categorization; and (3) attention. While these are ontic, metaphysical categories, these "psychological primitives" are eerily similar to Heidegger's ontological categories of mood, understanding, and fallenness. Mood or affect is conceptualized by Barrett as a process of evaluating how things matter, e.g., pleasant or unpleasant. Likewise, Heidegger's ontological category of *Befindlichkeit* identifies mood as revealing how things matter. The second primitive of categorization is parallel with Heidegger's notion of fallenness. To understand what things mean, we draw from the available meanings of culture and language, and these appear in perception as distinct categories of things or other people. Finally, attention is closely linked to the notion of understanding in Heidegger, since he identifies understanding as a process of projecting possibilities into the future. In the present moment at any given time, this activity translates psychologically into the orientation of attention to this or that object or person. Barrett suggests that neuroscience should dispense with seeking distinct anatomical structures for emotion and cognition, respectively, and instead identify how the brain, body, and world interact to produce our ongoing experience through the modulation of valuation, categorization, and attention. By doing so, she seems to be founding neuroscience on careful phenomenological analysis rather than the more arbitrary whims of folk wisdom.

From Heidegger to the Brain

From Heidegger and Barrett, we have seen that a phenomenological epoché allows the researcher to set aside or bracket assumptions about the metaphysical categories of emotion and cognition. The result is the ontological structure of experience as having the qualities of valuation, categorization, and attention, which map onto Heidegger's ontological categories of mood, understanding, and fallenness. When we examine the role of emotion in the brain, we find that what typically passes for "emotion" in neuroscience—and thus valuation or mood—tends to be associated with activity in the amygdala. And, furthermore, as we shall see, the activities of the amygdala are seen to be closely associated with functions related to understanding and fallenness in Heidegger's terminology and categorization and attention in Barrett's.

Research on the amygdala over the past decade or so has found that the activity of this almond-shaped structure in the limbic system is closely association with functions that include learning and memory, attention and perception, social behavior, and the regulation of emotion (Phelps & LeDoux, 2005). Phenomenologically, moods and emotions disclose how things matter to the person in terms of the past, which implies that affect should be closely associated with learning and memory. Neuroscientific evidence supports this contention in that activity of the amygdala is associated with enhancing the consolidation of memory, which means that highly relevant personal events that matter deeply will be remembered in more richly elaborated detail (McGaugh, 2000, 2004). Likewise, since mood is ontologically and structurally related to understanding, affects should also influence what present events are the most salient to our attention and how they matter. And in addition, affects should influence how we are anticipating future events, or using more Heideggerian terminology, moody understanding should be seen to reveal to us our possibilities for being-in-the-world. Again, these insights are supported by research on the brain, which finds, for example, that the emotional state of fear enhances future attunement to a feared phenomenon. As would be predicted by this hypothesis, activity in the amygdala has been found to be closely linked to plasticity in the auditory cortex as a result of exposure to fear-inducing sounds (Poremba & Gabriel, 2001; Quirk, Armony, & LeDoux, 1997).

In addition to influencing anticipation of future events, mood states should also modulate attention, directing the person to aspects of the phenomenal field that are of particular significance for the person. Again, neuroscience supports this contention. The amygdala has been found to be closely linked with sensory pathways in the cortex (Amaral, Behnieu, & Kelly, 2003); thus, intense states of affect predictably produce increased cortical attention and vigilance to personally relevant events in the world (Armony, Servan-Schreiber, Cohen, & LeDoux, 1997). Likewise, when a person has damage to the amygdala, he or she is found to lack the expected modulation of attention associated with affective states (Anderson & Phelps, 2001).

Heidegger's theory of mood likewise predicts that human affective responses should be especially attuned to social dynamics. This is partly what the existential category of fallenness alludes to—our tendency to understand ourselves in terms of our relationships to others and to appeal to culture as a way to make sense of our lives. Not surprisingly, then, affect and neural activity in the amygdala is closely linked to social behavior. Those with damage to the amygdala, for example, have much greater difficulty interpreting facial expressions (Adolphs et al., 1999) and tracking the eyes of other people in ways that are essential for the interpretation of mood (Adolphs et al., 2005).

While these fascinating associations between Heidegger and neuroscience are illuminating, they are only a hint of the potential for better understanding of experience if phenomenology and neuroscience were to be more seriously married in future investigations. Of particular importance is research that will focus on rich, experience-near descriptions of specific, momentary moods using a rigorous phenomenological method of analysis. Once invariant thematic descriptions are produced with the phenomenological method and validated experimentally, neuroscience can highlight the neural correlates of these experiences in ways that can be mutually beneficial. Neuroscience can gain by having more fine-grained descriptions to map onto the structure of the brain, and phenomenology will be refined to the extent that its findings are checked against the functional anatomy of the brain and nervous system. We have, then, a recipe for a fruitful neurophenomenology and the makings of a truly radical experiential revolution.

References

Adolphs, R., Gosselin, F., Buchanan, T. W., Tranel, D., Schyns, P., & Damasio, A. R. (2005). A mechanism for impaired fear recognition after amygdala damage. *Nature, 433*, 68–72.

Adolphs, R., Tranel, D., Hamann, S., Young, A.W., Calder, A. J., Phelps, E. A.,...Damasio, A. R. (1999). Recognition of facial emotion in nine individuals with bilateral amygdala damage. *Neuropsychologia, 37*, 1111–1117.

Amaral, D. G., Behniea, H., & Kelly, L. (2003). Topographic organization of projections from the amygdala to the visual cortex in the Macque monkey. *Neuroscience, 118*(4), 1099–1200.

Anderson, A. K., & Phelps, E. A. (2001). Lesions of the human amygdala impair enhanced perception of emotionally salient events. *Nature, 411*, 305–309.

Anderson, A. K., & Phelps, E. A. (2002). Is the human amygdala critical for the subjective experience of emotion? Evidence of intact dispositional affect in patients with amygdala lesions. *Journal of Cognitive Neuroscience, 14*(5), 709–720.

Armony, J. L., Servan-Schreiber, D., Cohen, J. D., & LeDoux, J. E. (1997). Computational modeling of emotion: Explorations through the anatomy and physiology of fear conditioning. *Trends in Cognitive Science, 1*, 28–34.

Baars, B. J. (1986). *The cognitive revolution in psychology*. New York, NY: The Guilford Press.

Banissy, M. J., Kanai, R., Walsh, V., & Rees, G. (2012). Inter-individual differences in empathy are reflected in human brain structure. *Neuroimage, 62*(3), 2034–2039.

Barr, S. M. (2006). *Modern physics and ancient faith*. Notre Dame, IN: University of Notre Dame Press.

Barrett, L. F. (2009). The future of psychology: Connecting mind to brain. *Perspectives on Psychological Science, 4*(4), 326–339.

Barrett, L. F., Bliss-Moreau, E., Duncan, S. L., Rauch, S. L., & Wright, C. I. (2007). The amygdala and the experience of affect. *Social Cognitive and Affective Neuroscience, 2*(2), 73–83.

Barrett, L. F., Mesquita, B., Ochsner, K. N., & Gross, J. J. (2007). The experience of emotion. *Annual Review of Psychology, 58*, 373–403.

Bitbol, M. (2003). A cure for metaphysical illusions: Kant, quantum mechanics, and the Madhyamaka (pp. 325–361). In B. A. Wallace (Ed.), *Buddhism and science: Breaking new ground*. New York, NY: Columbia University Press.

Bruner, J. S. (1961). *A study of thinking*. New York, NY: John Wiley and Sons, Inc.

Bruner, J. S. (1993). *Acts of meaning*. Cambridge, MA: Harvard University Press.

Bruner, J. S., Goodnow, J. J., & Austin, G. (1956). *A study of thinking*. New Brunswick, NJ: John Wiley and Sons Inc., Transaction Publishers.

Churchland, P. S. (1986). *Neurophilosophy*. Cambridge, MA: The MIT Press.

Clegg, J. W. (2009). The problems of reductionism in the social sciences. In J. W. Clegg (Ed.), *The observation of human systems: Lessons from the history of anti-reductionistic empirical psychology* (pp. 1–8). New Brunswick, NJ: Transaction Publishers.

David, N., Bewernick, B. H., Cohen, M. X., Newen, A., Lux, S., Fink, C. R.,...Vogeley, K. (2006). Neural representatives of self versus other: Visual-spatial perspective taking and agency in a virtual ball-tossing game. *Journal of Cognitive Neuroscience, 18*(6), 898–910.

Dreyfus, H. L. (1972). *What computers can't do: A critique of artificial reason*. New York, NY: Harper & Row.

Ellis, R. D. (2006). Phenomenology-friendly neuroscience: The return to Merleau-Ponty as psychologist. *Human Studies, 29*(1), 33–55.

Forgas, J. P. (2002). Affective influences on attitudes and judgments (pp. 596–618). In R. J. Davison, K. R. Scherer, & H. H. Goldsmith (Eds.), *Handbook of affective sciences*. New York, NY: Oxford University Press.

Gallagher, S. (1997). Mutual enlightenment: Recent phenomenology in cognitive sciences. *Journal of Consciousness Studies, 4*(3), 195–214.

Gallagher, S. (2007). Neurophilosophy and neurophenomenology. In L. Embree & T. Nelson (Eds.), *Phenomenology* (Vol. 5, pp. 293–316). Bucharest, Romania: Zeta Press.

Gallagher, S., & Sorensen, J. B. (2006). Experimenting with phenomenology. *Consciousness and Cognition, 15*, 119–134.

Gardner, H. E. (1987). *The mind's new science: A history of the cognitive revolution*. New York, NY: Basic Books.

Gendlin, E. T. (1978b). *Befindlichkeit*: Heidegger and the philosophy of psychology. *Review of Existential Psychology and Psychiatry, 16*(1–3), 43–71.

Heidegger, M. (1962). *Being and time*. New York, NY: HarperCollins.

Husserl, E. (1982). *Ideas pertaining to a pure phenomenology and to a phenomenological philosophy – First book: General introduction to a pure phenomenology* (F. Kersten, Trans.). The Hague: Nijhoff. (Original work published 1913)

Lambie, J. A., & Marcel, A. J. (2002). Consciousness and the varieties of emotion experience: A theoretical framework. *Psychological Review, 109*(2), 219–259.

Lazarus, R. S. (1994). *Emotion and adaptation*. New York, NY: Oxford University Press.

LeDoux, J. E. (1996). *The emotional brain: The mysterious underpinnings of emotional life*. New York, NY: Simon & Schuster.

LeDoux, J. E. (2000). Emotion circuits in the brain. *Annual Review of Neuroscience, 23*(1), 155–184.

Lutz, A., Lachaux, J.-P., Martinerie, J., & Varela, F. J. (2002). Guiding the study of brain dynamics using first-person data: Synchrony patterns correlate with on-going conscious states during a simple visual task. *Proceedings of the National Academy of Science USA, 99*, 1586–1591.

MacLean, P. D. (1949). Psychosomatic disease and the "visceral brain": Recent developments bearing on the Papez theory of emotion. *Psychosomatic Medicine, 11*, 338–353.

Mauss, I. B., Levenson, R.W., McCarter, L., Wilhelm, F. H., & Gross, J. J. (2005). The tie that binds? Coherence among emotion experience, behavior, and physiology. *Emotion, 5*, 175–190.

McGaugh, J. L. (2000). Memory – a century of consolidation. *Science, 287*, 248–251.

McGaugh, J. L. (2004). The amygdala modulates the consolidation of memories and emotionally arousing experiences. *Annual Review of Neuroscience, 27*, 1–28.

Merleau-Ponty, M. (1983). *Structure of behavior*. Pittsburgh, PA: Duquesne University Press. (Original work published 1947)

Merleau-Ponty, M. (2002). *Phenomenology of perception* (2nd ed.). New York, NY: Routledge.

Moriguchi, Y., Ohnishi, T., Mori, T., Matsuda, H., & Komaki, G. (2007). Changes of brain activity in the neural substrates for theory of mind during childhood and adolescence. *Psychiatry in Clinical Neuroscience, 61*(4), 355–363.

Mruk, C. J. (2008). The psychology of self-esteem: A potential common ground for humanistic positive psychology and positivistic positive psychology. *The Humanistic Psychologist, 36*(2), 143–158.

Noe, A. (2006). *Action in perception*. Cambridge, MA: The MIT Press.

Noe, A. (2010). *Out of our heads: Why you are not your brain, and other lessons from the biology of* consciousness. New York, NY: Hill & Wang.

Papez, J. W. (1937). A proposal mechanism of emotion. *Journal of Neuropsychiatry and Clinical Neuroscience, 7*(1), 103–112.

Petitot, J. (1992). *Physique du sens*. Paris, France: Editions du CNRS.

Phelps, E. A., & LeDoux, J. E. (2005). Contributions of the amygdala to emotion processing: From animal models to human behavior. *Neuron, 48*, 175–187.

Poremba, A., & Gabriel, M. (2001). Amygdalar efferents initiate auditory thalamic discriminative training-induced neuronal activity. *Journal of Neuroscience, 21*, 260–268.

Quirk, G. J., Armony, J. L., & LeDoux, J. E. (1997). Fear conditioning enhances difference temporal components of tone-evoked spike trains in auditory cortex and lateral amygdala. *Neuron, 19*, 613–624.

Robbins, B. D. (2003). *Joy and the politics of emotion: Towards a cultural therapeutics via phenomenology and critical theory*. Doctoral dissertation, Duquesne University, Pittsburgh, PA.

Robbins, B. D., & Fleming, R. (2008, August). *The unwanted exposure model of embarrassment: Experimental validation of an empirical, phenomenological study*. Poster presented at the 2nd Annual Conference of the Society for Humanistic Psychology, Division 32 of APA, Norwood, MA.

Robbins, B. D., & Parlavecchio, H. (2006). The unwanted exposure of the self: A phenomenological study of embarrassment. *The Humanistic Psychologist, 34*(4), 321–345.

Robbins, B. D., & Vandree, K. (2009). The self-regulation of humor expression: A mixed method, phenomenological investigation of suppressed laughter. *The Humanistic Psychologist, 37*(1), 49–78.

Sartre, J. P. (2003). *Sketch for a theory of the emotions* (2nd ed.). New York, NY: Routledge. (Original work published 1939)

Thom, R. (1975). *Structural stability and morphogenesis*. Reading, MA: W. A. Benjamin.

Thompson, E. (2007). *Mind in life: Biology, phenomenology, and the sciences of mind*. Cambridge, MA: Harvard University Press.

Uhlhaas, P. J., & Mishara, A. L. (2007). Perceptual anomalies in schizophrenia: Integrating phenomenology and cognitive neuroscience. *Schizophrenia Bulletin, 33*(1), 142–156.

Varela, F. J. (1996). Neurophenomenology: A methodological remedy for the hard problem. *Journal of Consciousness Studies, 3*, 330–350.

Varela, F. J., Thompson, E., & Rosch, E. (1991). *The embodied mind: Cognitive science and human experience*. Cambridge, MA: The MIT Press.

Vogeley, K., May, M., Ritzl, A., Falkai, P., Zilles, K., & Fink, G. R. (2004). Neural correlates of first person perspective as one constituent of human self-consciousness. *Journal of Cognitive Neuroscience, 16*(5), 817–827.

Wertz, F. J. (1983). Revolution in psychology: Case study of the new look school of perception. In A. Giorgi, A. Barton, & C. Maes (Eds.), *Duquesne studies in phenomenological psychology*. Pittsburgh, PA: Duquesne University Press.

Wertz, F. J. (Ed.). (1994). *The humanistic movement: Recovering the person in psychology*. New York, NY: Gardner Press.

Neurophenomenological Praxis: Its Applications to Learning and Pedagogy

Robert Garfield McInerney

> *Between the exploration and what it will teach me, between my movements and what I will touch, there must exist some relationship....*
>
> (Merleau-Ponty, 1968, p.133)
>
> *Teaching must be able to get caught up in the play of learning.*
>
> (Davis, Sumara, & Luce-Kapler, 2005, p. 148)

The purpose of this chapter is to first explore potential hybrid theories and methodologies that will help to explicate specific and immediate moments of learning, such as *situated* learning as well as *embodied* and *enactive* learning and, second, to advocate for the use of a pedagogical portfolio assessment and praxis that is appropriate for adult learners and that values these ways of learning.[1] Accordingly, in continuation of my previous work (McInerney, 2010), I will synthesize specific ways of learning (situated, embodied, enactive) with a neurophenomenologically inspired pedagogy and praxis for the purpose of liberating these ways of learning from educational subjugation (Fendler, 1998; Foucault, 1980; hooks, 1994; Kincheloe, 2008).[2]

To briefly define some of the terms mentioned above:

1. *Situated learning*. Sawyer and Greeno (2009) relate, "from a situated perspective, learning is the gradual appropriation, through guided participation, of the ability to participate in culturally defined, socially situated activities and practices" (p. 354; see also Greeno, 1998; Lave & Wenger, 1991; Walkerdine, 1997). For Lave and Wenger (1991), "learning is not merely situated in practice – as if

R.G. McInerney, Ph.D. (✉)
Department of Humanities and Human Sciences, Point Park University,
201 Wood Street, Pittsburgh, PA 15222, USA
e-mail: rmcinerney@pointpark.edu

S. Gordon (Ed.), *Neurophenomenology and Its Applications to Psychology*,
DOI 10.1007/978-1-4614-7239-1_2, © Springer Science+Business Media New York 2013

it were some independently reifiable process that just happened to be located somewhere; learning is an integral part of a generative social practice in the lived-in world" (p. 35).

2. *Embodied cognition and learning*. Embodied cognition is prereflective perceiving, thinking, and learning that emerges from the lived body in action (Gallagher, 2005). As Merleau-Ponty (1962/1989) states: "My body has its world, or understands its world, without having to make use of my 'symbolic' or 'objectifying' function" (pp. 140–141). When thoughts, choices, and attitudes are considered at one with the proprioceptive, sensorimotor, and perceptual body, this helps us understand a diversity of ways in which humans learn (see McInerney, 2010).
3. *Enactive cognition and learning*. Enactive cognition and learning suggests that perceiving, thinking, and learning are actively inseparable from our meaningful engagement with the environment (Bateson, 1972/2000; Fenwick, 2000; Varela, Thompson, & Rosch, 1991). Francisco Varela (1999) tells us that enaction includes "coupling of the cognitive agent, a permanent coping that is fundamentally mediated by sensorimotor activities" as well as "the autonomous activities of the agent whose identity is based on emerging, endogenous configurations (or self organizing patterns) of neural activity" (p. 272). Interpretively, enaction implies that learning happens by "…having a body with various sensorimotor capacities [and…] these individual sensorimotor capacities are themselves embedded in a more encompassing biological, psychological, and cultural context" (Varela et al., 1991, p. 173). It is important to note that these couplings, which I will show are describable, are not determining or causal of the active (learning) agent.
4. *Neurophenomenology*. Neurophenomenology began as a theory that emerged from the melding of neuroscience with phenomenology, which in turn shaped enaction: "Weaving together…the phenomenological and neurobiological, in order to bridge the gap between subjective experience and biology, defines the aim of neurophenomenology, an offshoot of the enactive approach" (Thompson, 2007, p. 15).

As a theory, neurophenomenology considers the interrelationship between neuroscience and phenomenology as potentially enlightening. It provides a plausible account of the need for naturalizing phenomenology as well as grounding neuroscience in the lived experiences of people. As Thompson (2007) suggests, neurophenomenology is also "experimental" when it "stresses the importance of collecting descriptive first-person reports of experience from phenomenologically trained subjects as a heuristic strategy for uncovering the physiological processes relevant to consciousness" (p. 338; see also Varela & Shear, 1999).

A pedagogical praxis is intended to liberate previously unrecognized styles of learning (e.g., situated, embodied, enactive) thus informing new ways of teaching (Cranton, 1996; Dirkx, 2002; Jarvis, 2005, 2006; Paulson & Paulson, 1994). The term "praxis" is distinguished here as research that puts into action a method for practical and liberatory results (Depraz, 1999; Kincheloe, 2005; Lather, 1986). Praxis also denotes a movement "beyond objectivism and relativism" toward an ongoing critical engagement with knowledge claims (Bernstein, 1983; see also Caputo, 1987; Ihde, 1986, 1993).

A portfolio assessment as described herein will be phenomenological in its approach as it seeks to qualitatively understand transformative learning (Mezirow, 1991). Transformative learning highlights what the adult learning agent interpretively brings to any given learning situation in order to transform previous knowledge paradigms (MacKeracher, 2004). A phenomenological portfolio assessment may emphasize the prereflective lived experiences of learning by gathering first-person, reflexive narratives of the "how" of situated, enactive, and embodied learning experiences in order to better understand the transformative dimensions related to the learning agent as intimately interrelated with the learning situation (see Jordi, 2011).[3] By using the terms "agent" and "agency," I am indicating the ability of the learner to actively and knowingly affect their world (see Burr, 2002; Gallagher & Zahavi, 2008; Martin & Sugarman, 1997). Agency, with regard to learning, is irreducible to any one theory or method and is irreplaceable in that each of us learns in our own way.[4]

Theorists have combined brain-based learning (i.e., insights from brain research) with education focused primarily on the child's development to enhance teaching and learning (Blackemore & Frith, 2005; Caine & Caine, 1991; Healey, 2004; Jensen, 1998, 2008; Sousa, 2010; Zull, 2002); however, studies have not utilized neurophenomenology, which I believe needs to be considered in the secondary education of, and research concerning the adult learner.

Learning, Cognitivism, and Cognitive Psychology

Learning, by its very nature, is a complex and fascinating phenomenon due to its associations with sociocultural norms, formal and informal educational methods, and psychology and neuroscience (Bruner, 1968, 1973, 1996; Gardner, 1983, 1991; Jarvis, 2006; Jensen 2008).[5] These diverse disciplines all have something to contribute to our knowledge about learning precisely because learning is an all-encompassing aspect of living (Illeris, 2008; Jarvis, 2005; Sawyer & Greeno, 2009). Consequently, there has been, and ought to be, many ways of studying the experience of learning using varying perspectives or techniques (Uttal, 2011).

Commonly understood, learning (to learn) means to gain knowledge of and to be informed (informed connotes inward change and formation). Learning can be understood as the collection of facts, instigating of change, and storing of information (Bruner, 1968). Learning may not have a fully formed beginning and end, and it may not be so easily separated from other cognitive activities – nor does learning seem to be, as we will see, a passive process of containment (Gardner, 1983; Maturana & Varela, 1987). Learning might best be described as diffuse (belonging to multiple contexts) whereby any learning experience can be understood as an embedded and emergent social phenomenon (Kincheloe, 2005; Lave & Wenger, 1991; Lemke, 1997; Vygotsky, 1978).

A resuscitation of all theories of learning is far beyond the scope and intent of this chapter. Sufficed to say, learning most certainly can be observed as a behavior. *Behaviorism* began with the premise that mental phenomena should be ignored in favor of overt observable physical actions (see Skinner, 1953, 1974; Uttal, 2011).[6] For behaviorists, the environment and the sensory and perceptual input of the organism were the developmental foundation for learning itself. Learning took place when the sensory and perceptual capacities of the organism responded to stimuli from the environment. Instances of learning, then, could be empirically observed and measured. Behaviorism also assumed that all behavior was rule-based and could be linked completely to causes and effects (Juarrero, 1999). Behaviorism contributed to our understanding of learning by demonstrating that some elementary forms of learning are instinctual (reflexive and associative) and the learner need not be explicitly aware of the responses that have been elicited.[7] Therefore, behaviorism suggests to educators that their main focus should be on environmental stimuli in the form of operant conditioning, rewards, punishments, and so on; as a result, the learner is, for the most part, passive (see Abraham, 2003). Behaviorism fits well with formal education and curricula, especially in the creation of lesson plans, objectives, goals, assessments, and standardizations. Davis et al. (2005) explain, "The problem is that behaviorism rests on the premise that the universe is mechanical and ultimately predictable – that is, complicated" (p. 59). Of most concern to this study, behaviorism isolated people from their everyday existence and presumed to understand learning. Even the supposedly straightforward reflex response must have a qualitative and meaningful dimension and so responses were not to "complex *situations*," but as if we could have "detached parts" responding to "isolated *stimuli*" (Merleau-Ponty, 1942/1983, p. 44). We can see that to understand the *complexity* of learning as enactive in situations, as well as embodied, would be beyond the behaviorist paradigm (Abraham, 2003; see also Merleau-Ponty, 1942/1983).

To further understand learning in relation to cognition, it is necessary to wonder about perception because as Merleau-Ponty (1964) believed, perception is the necessary beginning of all cognitions, affect, and intentions. Cognition is the process or activity of thinking (Neisser, 1976). Gestalt psychology, in part, is credited for widening our understanding of perception in relation to cognition and learning (Köhler, 1947). Gestalt denotes perceptional relationships that are inherently unified. Influenced by Brentano's act psychology, which stressed the holistic and cohesive interactions of the individual with their environment, Wolfgang Köhler (1938) showed that insight learning transcended simple stimulus and response models of learning; it demonstrated a creative, holistic, and synthetical ability on the part of the learning subject (Brennan, 1994; Uttal, 2011).[8] Perception must be more than input recognition of an object or simple response from a conditioned stimulus. Perception, for the Gestaltists, became indubitably tied to active and insightful cognitions, and they expanded our understanding of higher-order mental processes.

Learning is developmental, for example, Jean Piaget's (1975) constructivist epistemology outlined progressive capacities (or "domain-general" modules) within the developing child (Karmiloff-Smith, 1997, p. 7). Piaget's scrutinizing of the ontogenesis of developmental cognitive processes in children presumably demonstrated that there were basic processes of logical thought that emerged in the interaction between the child and the environment. Learning then, for Piaget, emerged from inborn and general schemas, but did not come to logical fruition until the child interacted with the environment in particular ways. Through the cognitive processes of assimilation and accommodation, schemas (see below) facilitated the construction of the real world that adults experience (as opposed to, e.g., the imaginative play world of the child). However, Piaget may have mistaken these logical processes as universal ontogenetic and rational development. Walkerdine (1984) explains that Piaget assumed *homo rationalis* was the pinnacle of development and thus "legitimate[s] and redirect[s] forms of classification of stages of development as regulatory and normalizing pedagogic practices" (pp. 176–177). Simms (1999) relates, "Piaget's bias is clear: the child will be an adult when he or she accepts that thinking is: (a) a matter of brain anatomy, (b) an internal event, (c) a subjective event, and (d) disconnected from the material world" (p. 303).

I will focus on specific, localized accounts of learning that may broaden our understanding beyond nativist or constructivist claims as well as instantiations of what is assumedly rational: "Piaget, however, as a theorist, never seems to have doubted the existence of a pregiven world and an independent knower with a pregiven logical endpoint for cognitive development" (Varela et al., 1991, p. 176).

Karmiloff-Smith (1997) points out, "For Piaget both gene expression and cognitive development are emergent products of a self-organizing system that is directly affected by its interaction with the environment" (p. 9; see also DeRobertis, 2011). This aspect of Piaget's work, as we shall soon see, fits well with the descriptive accounts of teaching and learning that I will provide. Furthermore, Piaget's constructivism, for some, is now critical constructivism, which avoids unwittingly privileging hegemonic ideologies influencing education and includes the role of the teacher as co-constructing the value of what is being learned (Giroux, 2001; hooks, 1994; Kincheloe, 2005).[9]

Early cognitive science postulated that perception is dependent upon mental organizing structures called *schemas* (Bartlett, 1932; Neisser, 1976). Classical cognitive philosophy, which was heavily influenced by René Descartes, imagined the mind as having and manipulating internal representations within a Cartesian theater of the mind (Dennett, 1991; Fodor, 1981). We witnessed the birth of *cognitivism* with Descartes' (1637/1980) focus on "precisely that part…that is a thing that thinks" (p. 76; see also Descombes, 2001; Dreyfus, 1972/1992) and the cogito that manipulates innate and internal ideas that have "their own true and immutable natures" (Descartes, 1637/1980, p. 85). According to cognitivism, the mind is primarily isolated from the living ecosystem (save for linear input/output); it is also disembodied, asocial, and ahistorical (Costall & Still, 1991; Descombes, 2001; Varela et al., 1991).

Cognitivism argued that perceiving, learning, and thinking can be explained as a reflective process that is marked by symbolically rule-governed, procedural operations within the mind (Descombes, 2001; Varela et al., 1991). The world was understood by cognitivism as a source of information (i.e., a series of inputs/outputs). Tim Ingold (2001) tells us that cognitivism made human beings into "devices for processing" knowledge as information (p. 114). Cognitivism began to imagine the cogito as existing somewhere within the mind as homunculi of all sorts: schemas, faculties, modules, and so on (Descombes, 2001). If the cogito works in such isolation and has its own self-evident and innate structure, then, as Descombes relates, "...the cognitive superstructures will all have to be explained by mechanical infrastructures" (p. 176; see also Dennett, 1991).

By the 1950s and 1960s, the cognitive movement began to hold sway, and information processing as learning (and thinking) became a truth that was taken for granted. The idea of schemas fit well with the emerging notions of learning as information processing or computationalism. Cognitive science relied heavily on imagining the mind as a computer that makes sequential and logical operations, and embraced the idea that the human brain is a *complicated* thinking thing or system (Winograd & Flores, 1987). Complicated systems are those that are highly structured and multifaceted (Cilliers, 1998; Davis et al., 2005). For example, the computer is complicated as it has many components with discrete functions that intercommunicate to form a sequential and logical series of events that ultimately drives the computer's representations and computations. Through analysis of these components and their interactions, we can understand all there is to know about the computer. This is partly because the computer lacks embodied experience, emotion, and mood; there is no worldly psychological life experienced by a computer (Dreyfus, 1972/1992, 1982; Winograd & Flores, 1987).[10]

Later, cognitive psychology posited that "perceiving is the basic cognitive activity out of which all others must emerge" (Neisser, 1976, p. 9). The preeminent cognitive psychologist Ulric Neisser explained that within the perceptual cycle, "schemata are anticipations, they are the medium by which the past affects the future" (p. 22). According to A. R. Luria (1976), perception "depends on historically established human practices...it possesses features that change along with historical development" (p. 21). Hence, schemas (or schemata) are "interactive," and as such can "be detached from the cycles from which they were originally embedded," and this leads to "imagining, planning, or intending" (Neisser, 1976, p. 23). Schemas begin to take shape, so to speak, the minute we are born and this ongoing process is inseparable from living itself. Perception, as Neisser believed, is a "constructive process" and complex schemata are full of anticipatory semantics (p. 20; see also Heidegger, 1926/1996). There is always already a perceptual "fore-structure" that prepares our seeing (and other senses) and allows us to recognize and conceptualize what we perceive (Heidegger). Merleau-Ponty (1962/1989) knew this: "...perception and the perceived necessarily have the same existential modality, since perception is inseparable from the consciousness which it has, or rather is, of reaching the thing itself" (p. 374).[11] The cognitive enactivists also understand perception as an anticipation of possible inputs instead of merely a causal reaction to inputs (Gallagher & Zahavi, 2008; Varela et al., 1991).[12]

By the 1980s, cognitive science and neuroscience had blended together to posit a more advanced computational mind. With progress in computer and brain sciences, the brain became an empirical puzzle that would eventually be solved (see Karmiloff-Smith, 1997). As Alva Noë (2009) argues, "Establishment neuroscience is committed to the Cartesian doctrine that there is a thing within us that thinks and feels. Where the neuroscientific establishment breaks with Descartes is in supposing that that thinking thing is in the brain" (p. 172).[13] While neuroscience gathered more evidence of localized function in the brain, the Cartesian mind acquiesced and became, perhaps, singularly reduced to the thing that thinks within the brain (Noë, 2009; Szasz, 1996; Varela et al., 1991). We can see then that an arduous task was taken on by the cognitivists, specifically, to fashion a structured mind that might eventually be placed in the brain (Wilson, 1998).

While we may include a cognitive neuroscientific account of the role the brain plays in preparing for and facilitating learning, such an account must avoid the pitfalls of cognitivism, namely, reductionism (see Notterman, 2000), rigid representationalism and computationalism (Clark, 1997, 2001; Ingold, 2001), and, in general, mentalism: "the premise that learning is a matter of building an internal model or representation of an external, pregiven reality" (Davis et al., 2005, p. 60). Further, our descriptions must take into account the emotions, moodedness, and embodiment of the learner in situations.

Perception is bodily and only with our human bodies (and brains) and actions (Noë, 2004) can we perceive the world (Gallagher, 2005; Merleau-Ponty, 1962/1989). Generally then, despite differences, we perceive like humans do, and this fact (*facticity*) facilitates a shared understanding of the world (see Heidegger, 1926/1996). Although we are not determined by the schemas in any strict sense, to be fair, schemas lead us into a good deal of conformity, and as such we tend to form culturally bounded, natural attitudes about life, learning, and living (Noë, 2009).[14] The aforesaid is an important part of our considerations in what follows.

In fact, for John Dewey (1938), "continuity," or the "experiential continuum," meant that the student brings to bear their previous learning on to future learning experiences (pp. 44–45). Dewey related that if we do not take into account the continuity and emergence of learning, then the "experience is treated as if it were something which goes on exclusively inside an individual's body and mind" (p. 39; see Gallagher, 2009; Koestenbaum, 1997).[15] The gradual shifting in education and pedagogy from strictly cognitivist (and nativist) to constructivist (Bruner, 1996; Kincheloe, 2005) has renewed interest in Dewey's work. Dewey foresaw the need to assess learning as immediately experienced. Dewey's account of learning mirrors cognitive ideas about the cyclical mode of perception as discussed above (see Dewey & Bentley, 1949; Neisser, 1976). For example, Neisser (1976) explains: "Perception and cognition are usually not just operations in the head, but transactions with the world. These transactions do not merely inform the perceiver, they also transform him" (p. 11). Dewey knew that to understand learning experiences, one must analyze the transactional relations between a student and their environment (Bateson, 1972/2000; Bredo, 1994; Clancey, 2008; Koestenbaum, 1997); likewise, Dewey and Bentley (1949) understood perception as a transactional relationship. This relationship happens in the midst of intention and meaning-making such that they each emerge in unison. Dewey recognized that through a complex involvement with others and the environment, learning could be analyzed beyond a simple stimulus/response model (i.e., behaviorism).

Dewey's advocacy for an experiential assessment of learning is commensurate with the educational movement toward situated learning (Lave & Wenger, 1991; Sawyer & Greeno, 2009) and constructivism (Poerksen, 2004). Dewey argued that the student is not in an objective situation and that "the conceptions of interaction and situation are inseparable from each other" (p. 43). While constructivism and education are indebted to Dewey (Vanderstraeten, 2002), theorists Lave and Wenger (1991) brought constructivism into particular learning community contexts.

Furthermore, Dewey's experiential description of the learning experience, generally, is an apt precursor to Varela (1996), Varela et al. (1991), and Maturana and Varela's (1987) analysis of learning as well as their depiction of experience as embodied and enactive (Bredo, 1994; Gallagher, 2009). And finally, we note that Dewey understood enactive perception long before the cognitive movement (see Gallagher & Zahavi, 2008).

We have seen that learning has been understood by behaviorism, mentalism, computationalism (cognitivism and information processing), as well as constructivism. What is discovered, or uncovered, about learning depends on ones' perspective and/or technique (see Gadamer, 1976; Heidegger, 1926/1996). Critical pedagogy (Freire, 1971; Kincheloe, 2008; Wink, 2010), existential learning theories (Jarvis, 2005), and critical constructivism (Kincheloe, 2005) have rightly pointed out the need for valuing the lived experience of learning while considering its embeddedness in sociopolitical constructions, gender and culture differences, power relations, and competing philosophical foundations (Breunig, 2005; Giroux, 2001; Kincheloe, 1991, 1999, 2005; Lather, 1991; Pinar, 2004).

We will look to enactivism whereby learning is a complex interrelationship and negotiation between the knower and the known (Bernstein, 1983; Davis et al., 2005; Maturana & Varela, 1987). Explained differently, we will see that the learning situation cocreates what the brain will do, and the brain actively engages and cocreates the learning situation. We will likewise consider the definition of learning broadly as meaning-making activity that accomplishes a goal, completes a task, or is a shift in one's perspective on life (Bateson, 1972/2000; Maturana & Varela, 1980; Varela et al., 1991). Davis et al. (2005) put forth that "learning is coming to be understood as a participation in the world, a co-evolution of knower and known that transforms both" (p. 64). By defining learning in these ways, we are opening the door, so to speak, for recognition of embodied, enactive, and situated learning as a "meaning-making journey of [adult] experiential learning" (Jordi, 2011, p. 195).

A method is needed that describes the immediate and present (en)active construction of learning while addressing the cutting-edge findings of cognitive neuroscience. Such a method should remain grounded in experience to recognize and value learning activity and skills that may be unrecognized and devalued in our formal educational system (Davis et al., 2005; Kincheloe, 2005; Sacks, 1999). When formal education too often adheres to the methodological positions of Cartesianism and cognitivism, teachers run the risk of ignoring other ways of learning (see Handley,

Sturdy, Fincham, & Clark, 2006; Kincheloe, 2005). Despite their liberating potential, these forms of learning (enactive, embodied, and situated) can be considered to be *subjugated.* By subjugated, I denote ways of knowing that have been hidden or devalued by our formal educational systems (see Foucault, 1980; McInerney, 2010).

The brain is perhaps the final frontier and ontological ground upon which formal brain-based education will likely continue to stake its claim (see Rose, 2005; Uttal, 2011). With this said, critical pedagogies and praxes must meet at these crossroads: neuroscience and phenomenology. And so, in the remainder of this work my task is threefold: first, to introduce neurophenomenology to psychologists and teachers; second, to use neurophenomenology to demonstrate how learning can be described as situated, embodied, and enactive; and third, to discuss how neurophenomenological praxis leads to a pedagogy that recognizes and liberates embodied, enactive, and situated learning.[16]

Phenomenology and Cognitive Neuroscience

Phenomenology is a specialized method of reflection and description that attempts to understand experience as it is immediately given, that is, not mediated by scientific constructs, axiomatic presuppositions, common sense, or experimentation (Gallagher & Zahavi, 2008; Husserl, 1913/1998).[17] Phenomenology wants to recognize the constitutive role of human consciousness within existence; how does the way we perceive the world as human beings contribute to the way we experience a phenomenon?

If we adhere to the idea that knowledge is simply passively received by the learner, then phenomenology will have little to say about learning. But, we will recognize how important phenomenology is to learning if we consider learning to be a phenomenon that happens differently in certain contexts (i.e., learning *extended* out to the world), or that we actively take part in (i.e., *enactive*), or that we are inextricably part of (i.e., *embedded*). Phenomenology tell us about how some experiences are born of the living body (i.e., *embodied*).

The reader will note the insinuation of cognitive terminology, parenthetically, in the above phenomenological interests regarding learning. That said, what is the cognitive connection with phenomenology? As Gallagher and Zahavi (2008) correctly point out, "Cognition is a secondary modification of our primary being-in-the-world, and it is only possible and attainable because we already are in the world" (p. 154). If we ignore existence and experience, and begin with only the cognitive domain or facts about the brain, we will have little more than abstractions that are devoid of a meaningful relatedness with our lives (Fisher, 1997; Winograd & Flores, 1987).[18] And, after all, what is learning if not a meaningful relatedness to the world? Therefore, we must ask if cognitive theories can find grounding in worldly and bodily everyday learning experiences.

The answer to the above is affirmative when we consider that phenomenology and some subdisciplines in cognitive science mutually reject cognitivism

(Clark, 1997; Costall & Still, 1991; Dreyfus, 1972/1992; Ingold, 2001; Varela et al., 1991). For example, Fred Wertz (1993) writes: "Phenomenology rejects cognitivism, the dogma that reality can only be experienced through cognitive constructs, whether they be units of information, neural networks, or schemas" (p. 20).[19] The extended, enactive, embedded, and embodied forms of learning mentioned previously are also based on this rejection of cognitivism (Osbeck, 2009).

Some of what cognitive science has postulated, such as dynamical systems (Juarrero, 1999; Thelen & Smith, 1994; van Gelder, 1998), embodied cognition (Gallagher, 2005), self-organizing systems (Maturana & Varela, 1980), enactive and distributed cognition (Pea, 1993), and complexity theory (Cilliers, 1998; Globus, 1995; Waldrop. 1992), all make sense with the ongoing work in cognitive neuroscience and phenomenology. Combining phenomenology with neuroscience leads us to the research method of neurophenomenology (Gallagher & Zahavi, 2008; Lutz & Thompson, 2003; Thompson, Lutz, & Cosmelli, 2005).

Neurophenomenology as Methodology

Epistemology, by its very nature, is conciliatory. What we know, how we have come to know it, its value, and whether there is more to know surely is a negotiation of theoretical backgrounds, expertise, reliability, and validity. In part, epistemology searches for the best methodology and method for understanding a phenomenon because epistemology is concerned with the relationship between the knower and the known (see Kincheloe, 1991). When an epistemological search settles on a method (i.e., applied epistemology), this is called methodology, namely, the philosophical and theoretical foundation that supports and validates a method (see Hoshmand & Martin, 1994). One must carefully consider methodology before the details of a method are laid out, or the method itself is used. My epistemological search has led me to consider learning as an act of intentional consciousness; in other words, as a unique meaning-making activity that is deeply interrelated with the proprioceptive-knowing body in action within particular situations. I suggest here a potentially productive negotiation between two parties: phenomenology and cognitive neuroscience. In an egalitarian and synergistic approach, both working together may provide some insight into learning experiences (Changeux & Ricoeur, 2000; Gallagher, 1997; Varela, 1996).

Following Francisco Varela (1996), neurophenomenology is defined as a way to understand first-person consciousness and lived experience through the use of phenomenological method and while relating the resulting discoveries of potential phenomenological invariants with third-person neurological findings. At first glance, neuroscience and phenomenology seem like an odd pairing. Admittedly a gloss, cognitive neuroscience values truth obtained deductively via objective methods; phenomenology values understanding obtained inductively through a subjective (and intersubjectively validated) method (Depraz, 1999). As with most dualisms held by theorists, when one or both sides begrudgingly (or not so begrudgingly) take the brave step toward the other, there can be a fruitful partnership and praxis (Changeux & Ricoeur, 2000; Depraz, 1999).[20] Some aspects of phenomenological research may be amenable to cognitive neuroscience, and vice versa, as long as we take heed that neither phenomenology nor neuroscience ought to give up crucial aspects of their methodology (Barclay, 2000; Ellis, 1999).

However, what do we make of the ontological differences between neuroscience and phenomenology? Phenomenologists reject the *naturalizing* of phenomenology that makes natural science *prior* to phenomenology and the sole method of generating truths about both the human condition and the world we live in (Clegg, 2006). On the other hand, phenomenologists may accept naturalizing that recognizes "that the phenomena it studies are part of nature and are therefore also open to empirical investigation" (Gallagher & Zahavi, 2008, p. 30).[21] Moreover, from a neurophenomenological perspective, there is an (inter)relationship between the brain and the world, and in some way or the other, the brain represents the world. If we concede that the brain, to some degree, represents the world, we have to add the caveat that this representation is not merely straightforward, easily objectified, or transparent (Clark, 1997; Ingold, 2001). True enough, phenomena, represented in human consciousness and the brain, are a part of nature, but the phenomenologist will insist that we co-constitute any phenomenon in question through the act of uniquely experiencing it (much less, studying it).

In fact, human consciousness' contribution to phenomena (i.e., acts of consciousness) was Edmund Husserl's (1913/1998) chief methodological concern, and the constituting acts of consciousness have been the starting place for neurophenomenology. This method must look beyond the Cartesian *I think* (i.e., beyond a study of consciousness) and additionally deal with, as Heidegger (1926/1996) and Merleau-Ponty (1962/1989) did, the Cartesian *I am*, namely, the person as embodied, intentional, mooded, and maintaining an active agency.

We may take seriously the oft quoted "mutual constraints" (Varela, 1996) and "mutual enlightenment" (Gallagher, 1997) affiliation that has been neurophenomenology's banner of sorts. Neurophenomenology is, at times, a rough road full of negotiations; but, it is an inroad nevertheless and one that may offer descriptive clues to the brain and to "learning environments as complex social systems" (Davis et al., 2005; Sawyer & Greeno, 2009, p. 354). With this beginning sense of methodology, let us now move on to phenomenology and neurophenomenology as a method and relate these methods to an informed pedagogical praxis.

Neurophenomenology as Method

The Epoché. Within the phenomenological tradition, what Husserl called the epoché is an attempt toward bracketing, or putting aside, the *natural attitude* (Husserl, 1913/1998; Spinelli, 1989; Zahavi, 2003). This bracketing is not easily done, for the natural attitude reveals that "I always find myself as someone who is perceiving, objectivating in memory or in phantasy, thinking, feeling, desiring"

(Husserl, 1913/1998, p. 54). We seem to assume, within the natural attitude, an "actuality" of existence (Husserl). This attitude makes existence factual, self-evident, and a "theory-independent reality" (Zahavi, 2003, p. 44) in which existence, or lived experience, is made into a concrete, taken-for-granted reality. According to the natural attitude, the real world is merely there; we do not construct it or co-constitute it—we simply perceive it accurately or inaccurately depending upon our perceptual capabilities (neurobiologically or interpretively).

If not for the epoché, the researcher would believe that a phenomenon could be explained simply through objective experimentation or reflection (i.e., gathering empirical data and applying statistical analysis and inference). In other words, bracketing does not lead to objectivity, quite the opposite; it leads to a recognition and intensification of subjectivity, which undercuts the object/subject dualism. For example, we assume that objects have properties that exist naturally within them and it is the accurate work of our sensory apparatus and reasoning that discovers the truth of these preexisting properties (Churchill & Wertz, 2001). Likewise, we stay within the natural attitude when we suppose we can discover the truth about people by examining their pre-given, innate, properties or the laws contained within genetics or the brain. We can see then, how the epoché helps the phenomenologist-researcher to avoid such diversions found within all that is posited through the natural attitude including subject/object dualism.

We may be tempted to think that the epoché, because of its relationship to the natural attitude, is bracketing *natural* science alone. However, this is not the case; Husserl (1913/1998) explains, "*All natural sciences and cultural sciences*, with their total stock of cognition, *undergo exclusion* precisely as sciences which require the natural attitude" (pp. 131–132). The epoché is ultimately a radical correction in our commonly held apperceptive experience in general; these experiences include our familial and sociocultural prejudices as well as the presuppositions that lie within the human sciences of sociology, psychology, and philosophy (see Moustakas, 1994). As Gallagher and Zahavi (2008) explain, "This realistic assumption [natural attitude] is so fundamental and deeply rooted that it is not only accepted by the positive sciences, it even permeates our daily pre-theoretical life…" (p. 22).

Note then that in terms of understanding the brain, we are not simply putting into abeyance the scientific explanations of the brain that have been postulated for centuries, but bracketing the prescientific and assumed natural folk wisdom about the brain as well. For example, that it must be the seat of the personality and learning, or that it must work like something encountered in nature, or mechanically like a clock, loom, piano, or computer (Gordon, 1988; Szasz, 1996).

Human perception is always alterable and, in fact, exists in a flow of alteration; thus, the epoché does not *transcend* human perception or even *my* subjective perception, but it does, or at least attempts to, transcend the natural attitude as apperception (Depraz, 1999; Moustakas, 1994). The epoché does not deny reality (Gallagher & Zahavi, 2008; Ihde, 1986); within phenomenological psychology it *transforms, radicalizes,* and prepares one's perception. There is no obscurity here, only discipline and practice (Ihde, 1986).[22] Bracketing is what the phenomenologist-researcher does to prepare herself or himself to be able to investigate a phenomenon. It is, in this way, no different than a researcher taking the time to set up safeguards for neutrality and objectivity in an experimental design.

When the phenomenologist-researcher begins the epoché, she or he creates a portal or entrance, if you will, in which to gain access to experiences as lived in the moment (Zahavi, 2003). As Depraz (1999) elucidates, the phenomenologist does not transform perceptions in isolation; there is an "intersubjective sharing of the reductive experience" (p. 105). Depraz inserts intersubjectivity, not in a member-checking validation toward the end of a qualitative study, but directly into the beginning stance and disciplined method of the phenomenologist (see also Lutz & Thompson, 2003). The epoché may be performed over a period of time, and in dialog with others.

An effective procedure in which to perform the epoché is to keep an ongoing autoethnography (Cho & Trent, 2006), or what Maso (2003) calls a "why interview." When the phenomenologist asks why take for granted this way in which to understand a phenomenon (i.e., questioning methodology) and asks why use such a method, she or he begins the epoché. However, more is needed as the researcher must explore his or her personal desires regarding the phenomenon (Maso).

Finally, we will keep in mind that the epoché cannot entirely wipe clean our presuppositions (Merleau-Ponty, 1962/1989). But, the researcher will eventually feel that she or he is sufficiently prepared and embark upon a systematic and thorough analysis of the interrelationship between the constituting structures of consciousness and how the *givenness* of the phenomenon ensues (Giorgi, 1975, 1997; Merleau-Ponty, 1962/1989). Givenness refers to the potential aftereffect of the epoché that has cleared a path, as best as possible, and in turn allows the phenomenon to emerge as presented in a new way (Churchill & Wertz, 2001).

The Phenomenological Reduction. An additional aspect of the phenomenological method is called the "phenomenological reduction," which involves richly descriptive accounts of the givenness of experience (Churchill & Wertz, 2001; Giorgi, 1975; Moustakas, 1994). It is important to note here that "reduction" does not mean to condense down to some elemental form, but to return to immediate apperceptions that have existed before the natural attitude. In fact, when the phenomenologist returns to experience as immediately given, it is complexity that is discovered, not minimalism. Often phenomenological descriptions seem poetic. Rightly so, for once the phenomenologist-researcher moves away from the natural attitude, their language becomes less wedded to objectified criteria and all that is prosaic or hackneyed in matter-of-fact explanations. The phenomenologist's poetic language should not obfuscate; rather, the descriptions should offer an improved acumen, intensity, and fidelity to shared experiences.

The researcher must stay with the act of perceiving an object or event, and if the object calls the phenomenologist away from this act, she or he must return to perception as given (Depraz, 1999; Finlay, 2009; Gallagher & Zahavi, 2008). Every time the researcher-phenomenologist takes for granted that the object of study is plainly this or that, the assumption must be called into question. Again, to demystify this method, recall that the epoché has put aside some presuppositions that have not been immediately given by the thing itself (or person), but given by the natural attitude and our social constructions generally (Berger & Luckmann, 1966; Gergen, 1995). Therefore, as the researcher performs the phenomenological reduction, she or he invariably comes to basic sensory descriptions that are prior to ad hoc conceptualizations. By the aforesaid, it is not meant that the reduction leads to raw sensory input, which presumably would be lacking in meaning. Using the reduction, the researcher moves away from what has been given about the phenomenon and toward the act of meaningful construction of the phenomenon (Dreyfus, 1982). Using the phenomenological reduction, researchers may expect to get to a language that has *more* fidelity to the phenomenon and a socially constructed perspective *less* indentured to the natural attitude (Ihde, 1993).

The reduction includes avoiding making aspects of experience better or more primary; it *horizontalizes* all experience (putting all experiences on one level; Husserl, 1913/1998). As Spinelli (1989) explains, "phenomenologists urge us to treat each bit of initial experience as if we have been given the task of piecing together some gigantic jigsaw puzzle without the prior knowledge of what image the completed puzzle depicts" (p. 19). The point of horizontalization is not to deny the reality that some things, especially in terms of potential danger, may be more important than others, it is to avoid over-conceptualizations that get in the way of the experience.

Imaginative free variations (*eidetic reduction*), which are part of the overall method of phenomenological reduction, remove that which is not essential to the phenomenon being studied and ask (explicitly, meticulously, and methodically) what differentiates the phenomenon being studied from other phenomena (Husserl, 1913/1998, pp. 147–164; Ihde, 1986). For Ihde, "the use of variations require obtaining as many *sufficient* examples or variations upon examples as might be necessary to discover the structural features being sought" (p. 40).

The phenomenological reduction ought to include a constant attempt to assure that the descriptive accounts presented remain grounded in the experience itself and, as such, not solely formulated from already existing theories. As Petitmengin and Bitbol (2009) conclude, "…becoming reflectively conscious of one's experience and describing it is a process which does not consist in observing or reflecting upon a pre-existing experience, but in an unfolding of experience elicited by precise acts" (p. 400). Further, every phenomenological description that appears out of the study of the phenomenon must be intersubjectively verifiable (Gallagher & Zahavi, 2008). According to Ihde (1986), "Intersubjective phenomenology is necessarily interdisciplinary phenomenology" (p. 133), and so to continue understanding any phenomenon will require a triangulation of methods.[23] Methodological triangulation simply shows that the researcher employs many (at least three) ways in which to understand the phenomenon (c.f. Robbins, 2006). These accounts, or protocols, can be analyzed using different types of phenomenological analyses and content analysis (Moustakas, 1994); for example, the reports may be reduced to "meaning units" and use free imaginative variation, so that invariant and essential structures may be found (see Giorgi, 1975, 1997).

With the aforesaid in mind, third-person objective studies of the brain are used in neurophenomenology. These can be obtained via fMRI and PET scans (Lutz & Thompson, 2003).[24] However, without narrative phenomenological descriptions (van Manen, 1990), we run the risk of leaping ahead of our phenomenon of study and prematurely providing theoretical and presumed essential features of the phenomenon (see Uttal, 2011).[25] As Wertz (1993) makes clear, "In phenomenological psychology, one starts with description and only resorts to construction after extensive intentional analyses have established first principles and fundaments of knowledge in the discipline" (p. 22).

Wertz's contention is of paramount importance in this study. Phenomenological data in the form of observations and first-person, reflective narratives may be front-loaded (Gallagher, 2003) into third-person methods. Front loading denotes having phenomenological insights drive a particular experimental design. As Gallagher and Sørensen (2006) explain:

> Just as experimental designs can be informed by specific theories, experiments can also be informed by phenomenological insights—that is, insights developed in independently conducted phenomenological analyses, or in previous neurophenomenological experiments. In such cases phenomenology is 'front-loaded' into the experimental design. (p. 125)

Ultimately, to explicate situated, enactive, and embodied learning, we may train participants in first-person phenomenological reflection, provide phenomenological observations using the phenomenological method, and front-load this data into an experimental (neurological) design.

Neurophenomenological Praxis

If people were merely complicated, our learning and our ways of teaching could easily be standardized. But, we are not complicated; we are complex. *Complexity* is a concept coming from, in part, current enactive and dynamic cognitive science. Complexity means that "the interaction between the system and its environment, are of such a nature that the system as a whole cannot be fully understood simply by analyzing its components" (Cilliers, 1998, p. viii; Waldrop. 1992). Complex systems are usually organic and are delineated by an interaction with the environment that is difficult to define simply as an enclosed input/output system. Cilliers (1998) tells us, "A complex system cannot be reduced to a collection of its basic constituents, not because the system is not constituted by them, but because too much of the relational information gets lost in the process" (p. 10).

The method of neurophenomenology, in part, seeks to uncover the characteristic features of *autopoiesis*. Autopoiesis denotes self-creation and self-organization of living systems (note that language is a complex and autopoietic system). Autopoietic systems transform and self-organize because of recurrent interactions with other dynamic, complex systems. Autopoiesis is a process that maintains and constitutes the system's unity (Maturana & Varela, 1980; Rudrauf et al., 2003). An autopoietic system is marked by multifarious connections that are transient and are temporally adhered to, or clustered with, other systems as they interact with each other in such a way as to continually produce and maintain the system's interrelationships (Maturana & Varela, 1987; Thompson et al., 2005).

The theory of autopoiesis can tell us how a complex system evolves and learns.[26] The relationship between self-organizing systems (autopoietic systems) and complexity is as follows: "The capacity for self-organization is a property of complex systems which enables them to develop or change internal structure spontaneously and adaptively in order to cope with, or manipulate, their environment" (Cilliers, 1998, p. 90).

Here our interest is basically concerned with two autopoietic systems. First, we can circumscribe the lived experiences of learning (situated, enactive, embodied) as a complex, autopoietic system. Second, we can understand the brain as an autopoietic system deeply interrelated with the lived experiences of learning as situated, embodied, and enactive. If we combine the method of neurophenomenology with an understanding of dynamic systems theory, we may outline autopoiesis and its multifarious connections as follows:

1. *Operational closure*, which designates that internal operations of the brain work in such a way that the by-product of its development remains within the neurobiological processes in the brain (Maturana & Varela, 1980). Notice that to say that the brain is operationally closed is not to say that it is isolated from the external world; instead, neurophenomenology argues that the brain is in harmony with its surrounding environment. By *closure* we note that the human brain is endogenous; the brain is openly in synchronization with the world. Juarrero (1999) relates, "Over time, that is, both phylogenetically and developmentally, people establish interdependencies between the environment and their internal dynamics such that the formal becomes part of their external structure: their boundary conditions" (p. 197).
2. *Structural coupling*, which indicates the observance of two or more autopoietic systems (people with other people, people within the ecosystem) that experience a reciprocally constituting interrelationship. Structural coupling denotes that "Two or more systems are coupled when the conduct of each is a function of the conduct of the other" (Thompson, 2007, p. 45).
3. *Attractors*. We can observe this conduct when we look for attractors, that is, that which is in the environment that draws the agent toward it. Juarrero (1999) tell us, "Attractors therefore represent a dynamical system's organization, including its external structure or boundary conditions" (pp. 152–153). Note that the attractor is part of this self-creating immediacy of autopoiesis because the attractor emerges uniquely within the situation and in a semiotic relation to the agent.
4. *Perturbations*, or triggers, actuate, but do not determine changes in consciousness and behavior (Thompson et al., 2005). The human brain is unintelligible without serious consideration of the way it is always already primed and expectant of any triggering stimuli (Gallagher & Zahavi, 2008; Thompson et al., 2005). Perturbations are also part of the autopoietic structural coupling, and their role in the enaction and experience of learning can be observed. A perturbation then is

anything in the environment that triggers, but does not necessarily determine reactions from the agent (Maturana & Varela, 1987). An *affective* perturbation is any trigger that is emotionally imbued and affects the emotional outlook one is in (see McInerney, 2010; Rietveld, 2008). Again, the perturbation does not act causally; it does not determine action or thought. A perturbation is observable, and only takes part in the inception of action and thought (Cox & Smitsman, 2008; Maturana & Varela, 1980).

5. *Affordances*. We also can observe the performance of structural coupling by noting emerging affordances within the self-creating immediacy of learning situations (Costall, 1995; Gibson, 1979; Good, 2007). J. J. Gibson (1979) described affordances as things in the environment that emerge as potentially useful beyond the assumed or normatively intended use of the thing; using the affordance *affords* an action or accomplishment. It is important to note that the affordance is considered as existing in an implicit, tacit, or liminal place in the environment (Gibbs & Van Orden, 2003; Polanyi, 1966; Rietveld, 2008). An affordance then is anything in the learning situation that by virtue of its interaction with the learning agent, and the demands of the situation, may transform to something that facilitates coping, adaption, and, generally, learning. Again, the affordance can be seen as emerging within the semiotic field of attention and intention and in relation to a previous history of structural coupling (Juarrero, 1999; Thibault, 2004).
6. *Trajectories*. The learner is attracted to the affordance by virtue of their previous trajectories. Thibault (2004) explains, "A trajectory is a persistence-in-time that arises through the organization of processes" (p. 4). A trajectory is a line of action or attraction that can be located and traced by an observer.

A *semiotic* trajectory is when autopoiesis produces a signification process, or meaning-making activity. Thibault (2004) explains that a "trajectory is a self-organizing system" because it is traceable to consciousness and the self, but without making either a concrete and completely knowable entity (p. 182). Trajectories are autopoietic because they are distinctively situated in and emerge from the complex interrelations of open systems; thus, any trajectory is self-creating and self-assembling in relation to open dynamical systems that structurally couple. The trajectory is the direction one learner takes in order to complete a task, move on to a new task, or bring into significance any sort of learning possibility (Thibault).

The semiotic trajectory, because of its meaningfulness to the learning situation and the learning agent in the situation, can be helpful in recognizing experientially situated, enactive, and embodied learning experiences. As Juarrero (1999) states, "Explaining why the agent took this path rather than that after forming the prior intention will require reconstructing the agent's background, circumstances, particular frame of mind, and reasoning, whether self-conscious or not" (p. 227). In fact, Juarrero tells us that when trajectories come together in a typical pattern, we observe the person's attractors within the socio-ecosystem.

Now, complex systems are dynamic and self-organizing (Juarrero, 1999) and as such maintain "structural congruence" with events, things, and others; this means we can follow, or describe, ontogeny as "the history of structural changes in a particular living being" (Maturana & Varela, 1987, p. 95). And yet, we can certainly agree that human beings are distinct from their eco-social systems and "operationally independent" (p. 95). In other words, while the process of the brain's organization (i.e., autopoiesis) is unchanging, its structure is ever changing during bottom-up, global synaptic transmission (Rudrauf et al., 2003). Remember that if we define learning as meaning-making activity, then meaning-making is "a distributed activity between body-brain systems and their ecosocial environments on diverse scalar levels of spatio-temporal and semiotic organization" (Thibault, 2004, p. 316). The "distributed activity" of meaning-making as learning may be described, but not reduced and concretized. Juarrero (1999) relates, "It is important to keep in mind that many complex systems and certainly the human neurological system are describable only by a manifold of mindboggling dimensionality" (p. 154).

How can "teachers as researchers" (Kincheloe, 1991) systematically observe autopoietic learning in action? As Kincheloe (2005) puts it, "Critical constructivism wants to return the sanctity of autopoiesis to the scholarly act, to pedagogy" (p. 109). I believe that teachers will both enhance their teaching, as well as advance our understanding of the brain and different learning experiences, if they set up deliberate praxes as an additive to their regular curriculum. Pedagogy becomes a collaborative praxis, which is then emancipatory (Greene, 1995; hooks, 1994). Therefore, let us now examine some praxes of learning (i.e., teaching strategies and examples) and move toward a pedagogical assessment that is informed by neurophenomenology as well as situated, embodied, and enactive learning.

Neurophenomenological Pedagogical Praxes

A Social Construction Pedagogical Praxis. I teach social psychology and social constructionism to undergraduates.[27] During class we do the following exercise: I ask students to take out a blank piece of paper and to fold the paper so that they can tear it into six pieces. Once the students have six, small, blank pieces of paper in front of them, I ask them to think about six of the absolute most important people, ideas, things, or events in their lives. I compel the students to wonder about these in relation to how they have come to be who they are. And so, each student proceeds to put one thing, person, event, or idea on one of six pieces of paper. I then ask them to fold the pieces of paper in half so they cannot see the response. Once this is complete, I ask the students to do the following: (1) allow the person next to you to randomly take one away; (2) without knowing what is on the paper, tear one up; (3) knowingly choose one and put it at arm's length; (4) knowingly share one with the person next to you; and (5) knowingly give one of your six pieces of paper to the person next to you.

As we do each aspect of the exercise, we imaginatively discuss the implications of each action taken in terms of the degree to which these people, things, ideas, and events socially construct us. Caine and Caine (1991), writing on brain-based learning, note that this "orchestrated immersion" of "dead" content into lively context makes the content metaphorical and thematic to the students' lives and at once concretely felt and experienced. The students ask, "Who are we without these people or events?" and "How our identities would have changed?" and, conceivably, "How would these losses and gains have come about?"

What this praxis does is quite fascinating: in essence, the students were asked to reflect upon previous attractors and trajectories, but we have brought them to life, if you will, in the form of current perturbations. We will remember that a perturbation is anything in the surrounding socio-ecosystem that triggers action or thought (Maturana & Varela, 1987).

Each part of the exercise is an affective perturbation, which is an emotional trigger that brings forth thought that is emergent from, and embedded in the situation. Having a significant person in one's life randomly removed, for example, creates an affective or emotional perturbation (Rietveld, 2008) that in turn provides increased reflexive focus on the learning experience (see McInerney, 2010; Paré, Collins, & Pelletier, 2002). The students in this example are more or less ready for these perturbations (or rules of this exercise) because of preceding experiences, or previous action-effects (Cox & Smitsman, 2008; Rietveld, 2008).

Emotion, far from being simply an impediment to learning and thinking, gives the learner insight into their actions. According to Freeman (1999), "We can begin to make sense of emotions by identifying them with the intention to act, and then to note their increasing levels of complexity" (p. 125). This praxis, consequently, has generated an affective-dispositional intentionality and awareness (Freeman, 2000; Lemke, 1997; Lewis & Todd, 2005; Rietveld, 2008), as well as prereflective and embodied skillful adaption and accomplishment (Gallagher, 2005; Merleau-Ponty, 1962/1989). In this praxis, each student expresses their adaption to the changes that take place when the six events, people, things, or ideas are altered. Language, people, and the environment (milieu) are "semiotic resources" and are available for learning as meaning-making and "in relation to the architecture and dynamical processes of the body-brain complex" (Thibault, 2004, p. 236).

When in the dynamic and situated circumstances of this exercise, students learn at the edge, or Vygotsky's "zone of proximal development," (see Daniels, 2008) of what they might know. Perturbations, I believe, help us to understand this learning experience beyond representational theory:

> Dynamic-system explanations focus on the internal and external forces that shape such trajectories as they unfold in time. Inputs are described as perturbations to the system's intrinsic dynamics, rather than as instructions to be followed, and internal states are described as self-organizing compensations triggered by perturbations, rather than as representations of external states of affairs. (Thompson, 2007, p. 11)

Moreover, the reflections upon the experience that are shared with the other students provide describable trajectories leading to attractors that, in turn, help the teacher-researcher to better understand the student's unique learning style.[28]

As Rietveld (2008) outlines, the skillful coping and prereflective bodily engagement with the world that Merleau-Ponty describes are comparable to the self-organization of brain and behavior. The adult brain, especially in terms of its higher-order perceiving, learning, and thinking, remains neurally plastic (Caine & Caine, 1991; Gross, 2000; Hill, 2001). From a dynamic systems perspective, the brain then begins a synchronizing neural appraisal of perturbations that are mediated by the autopoietic coupling (internal) with the external situation (Varela et al., 1991; Varela & Thompson, 2003). It is likely that the brain's cortical and subcortical areas become more actively engaged in a way that they would not be if, as teachers, we adhered to Cartesian and cognitivist paradigms of pedagogy (see Lewis & Todd, 2005). When the students are in the situation of this social construction praxis, there is, theoretically, a phase synchronization involving the prefrontal lobe, the limbic system, and the brain stem (Lewis & Todd; Varela, Lachaux, Rodriguez, & Martinerie, 2001).

Sabotage as Pedagogical Praxis. *Sabotage* is a naturalistic environment teaching strategy in which the teacher as researcher (Kincheloe, 1991) sets up a learning situation so that something the learner wants is in the immediate environment as a possibility, but potentially unattainable.[29] This is often done with young children, or children with learning differences. However, the teaching strategy can be performed with adult learners in the form of improvising praxes.

In one of my classes, I ask students to tell a story to each other using limited language: only a few gestures are allowed and mostly props; therefore, the props come into view as an *affordance*. In this praxis, the teacher-researcher may follow the attraction a particular student demonstrates for a particular prop and a particular *line of action* in using the prop. As the students struggle to communicate with each other through a good deal of laughter, they come to realize the nuances of interpersonal communication and the role of *signifiers* (not necessarily formalized language) in the production of meaning. If a student wants to continue the story, the student is attracted to certain props, which in turn emerge as affordances. An attractor, although certainly related to a perturbation (see above), does not impinge upon the learner but, instead, draws the learner in toward it. This *drawing in* toward the affordance is meaningful (a "semiogenetic trajectory") in a way that the learner is likely not completely aware of (Thibault, 2004). Likewise, the trajectory is bound to the attractor, but the trajectory is the traceable action or gesture toward the attractor, which may, or may not emerge as an affordance.

Thibault (2004) relates: "interpersonal meaning orients interactants in terms of the given phenomenon's value-laden salience for action" (p. 212). In the pedagogical praxis described above, the learning agent will only be attracted to certain props that may allow an affordance; this attraction is, in some ways, unique to that learner. Put simply, the prop does not only trigger or act upon the learner; rather, the learner is drawn to the prop based upon previous experiences and, theoretically, previous trace-synaptic clustering and strengths of neural connections ("weights," according to connectionism; see Globus, 1995).[30]

The sabotage, then, is the limitation deliberately constructed within the environment by the teacher-researcher (i.e., only allowing props). Overall, this meaning-making activity as learning emerges not from a preplanned schemata or Cartesian cogito, but from a complex array of affordances, trajectories, perturbations, and in relation to the autopoietic closed system.

Actions (and "action learning"; Jarvis, 2005) are embodied in this exercise; "… the biological body is not a structure through which one learns, but a structure that learns" (Davis et al., 2005, p. 66). Each time a prop is used, the teacher may note the autonomous learning agent's adaption within the complexity of the learning situation. The embodied learner becomes the learning body, which molds itself to the prop at hand, fits itself into the situation, and stretches to communicate (see Gallagher, 2005; McInerney, 2010).

We can additionally witness the role of emotion and mood in this learning experience (see Dirkx, 2002; Goleman, 1995; LeDoux, 1996). As my students report, this praxis is viscerally felt. The sabotage obstructs habitual patterns of interpersonal relations; it obstructs the students' deliberate intention and plans. In terms of the emotional brain, "…it may be during states of obstructed and extended intentionality that emotions become the object of explicit awareness and refine present intentions or establish intentions of their own" (Lewis & Todd, 2005, p. 219; see also Freeman, 2000; Gibbs & van Orden, 2003). Because each prop represents the students' desire to communicate, the prop-perturbation acts as an affective obstruction and compels the students to rethink and adapt to the changing dynamics of language, signification, and communication. The lesson plan, then, is to teach a sophisticated understanding of communication beyond direct transmission of information.

The Identity Game. In one of my classes, we play what I call the identity game. Many student volunteers come up before the class and sit in chairs approximately 2 ft apart. I then ask each student to take turns and identify herself or himself using any sort of label or experience including things like gender, ethnicity, race, occupation, likes or dislikes, and habits. One of the simpler objectives of the identity game is to get students to move their bodies, which, of course, in turn facilitates thinking and learning in terms of embodiment (Sheets-Johnstone, 1990).

Lave and Wenger (1991) tell us that "A person's intentions to learn are engaged and the meaning of learning is configured through the process of becoming a full participant in a sociocultural practice" (p. 29). As the volunteers identify each other, the rest of the students respond as to whether they believe the identification is *relational* (i.e., the meaning of the identity is born of a particular interpersonal relation), *positional* (i.e., the meaning of the identity is born of some social position, usually connected with power and hierarchy), or *contextual* (i.e., the meaning of the identity seems to be more about the context).

For Varela et al. (1991), what they refer to as "context-dependent know-how" is the "essence of creative cognition" (p. 148). It is this know-how that emerges when we follow the learner as situated. The learners' abilities are, as Varela et al. say, "rooted in the structures of our biological embodiment, but are lived and experienced within a domain of consensual action and cultural history" (p. 149).

And so, I ask students to physically move either closer or farther away depending on our interpretation of the identity; in other words, do the students identify with each other, and if so, how is this identification represented in terms of their physical proximity, sense of closeness, and community? The point here is to *feel* the *enacting* of interpersonal connectivity and community. The students are put in a situation of learning that is both participatory and social (Handley et al., 2006; Lave, 1988; Lave & Wenger, 1991; Siegel, 1999). Autopoiesis thus becomes a living social process (Ingold, 2001; Luhmann, 1990; Toren, 2001).

As outlined above, Maturana and Varela (1987), interested in the biological and neurological maintenance of identity, describe structural coupling as observable interrelationships between two or more autopoietic systems (people, students) and the surrounding socio-ecosystems that have not been preprogrammed and therefore cannot be predicted. They outline social, *third-order coupling*, in which a given person's unique ontogeny pairs with others creating "co-ontogenies," which then form "third-order unities" (p. 193). Here "ontogeny" refers to the diverse origin and historical development of people (Maturana & Varela). Third-order structural coupling then becomes a social phenomenon of communication: "the coordinated behaviors mutually triggered among the members of a social unity" (p. 183). Note that the unique learner is not lost in this; she or he brings to the exercise a unique ontogenic history, which is, in part, why there can be infinite variation in this simple demonstration.

Because of the complexity (see above) of our development, and our privacy and individuality, the teacher-researcher may observe ever "new dimensions of structural coupling" (Maturana & Varela, 1987, p. 176). This exercise brings forth a fascinating tension between privacy and sociality, the student's sense of intersubjectivity, and an excellent example of "distributed cognition" (Hutchins, 1995) in which thinking is distributed among the things and people of a distinct and complex situation.

Captivatingly, each student becomes an *attractor* of the other when they reveal themselves. Once attracted to a particular description of identity, the student moves toward the other. For Gallagher (2005) we understand others in an "embodied practice" and "…in most intersubjective situations we have a direct understanding of another person's intentions because their intentions are explicitly expressed in their embodied actions, and mirrored in our own capabilities for action" (p. 224). If we accept learning as an enactive social phenomenon that can be set up, if you will by the teacher, then we will do well to describe learning as attentional and intentional in the learning situations. We can know when someone has learned based on their responses or as actions observed. In what follows, we will recognize the learner's attention and follow their intention because perception *as* attention and action *as* intention are hermeneutically cyclical (see Gibson, 1979; Juarrero, 1999) and can be observed and described. According to Ingold (2001), Gibson's (1979) "education of attention" shows us that we learn to attend through a "fine-tuning or sensitization of the entire perceptual system, comprising the brain and peripheral receptor organs along with their neural and muscular linkages, to particular features of the environment" (p. 142). Theoretically, there is an inseparability of intention and attention: thus to observe and interpret one's intentions is to interpret meaning and experience in relation to one's perceptual and apperceptual attention (Merleau-Ponty, 1962/1989). Cox and Smitsman (2008) argue that our intentions and observable choices "emerge in a self-organizing way from the coupled dynamics of all contributing subsystems" (p. 329).[31]

Thus far we have looked at pedagogical examples regarding society and community as well as intra- and interpersonal relations. What can we say about embodied experiences in relation to neurophenomenology and pedagogy?

Saliva or Spit? In another class, I ask for brave student volunteers to spit, or let their saliva drip into a small cup. Of course, this act is considered gross by the students. But it gets worse; I then ask the volunteers to drink their own saliva! After much uneasiness and laughter, most of the students invariably refuse and our discussion begins.

Although one might think that physiology holds the essential truth of saliva as something objectively present, this is insufficient. In fact, saliva understood this way would be the purview of science in terms of reality (what is assumedly really real), but we are interested in existence, that is, how saliva is experienced as meaningful. Meaning, as we discuss in class, is generated socially and interpersonally. The saliva becomes spit when it is out of the mouth and out of the mouth in certain contexts.

Husserl viewed all human consciousness as intentional (Husserl, 1913/1998), that is, there is always already some direction and intent, some implication of, and toward the world within consciousness. Caputo (1987) believes that Husserlian intentionality *is* interpretation because intentionality is a prestructure (i.e., a pre-intention) for the possibility of human understanding and experience. We bring to all experiences a *fore-structure* of understanding, which has been, for the most part, socioculturally constructed (Caputo; Heidegger, 1926/1996). Therefore, the intentional object, spit or saliva, makes sense only by way of a relational hermeneutics—socially interpreting together. Larkin, Eatough, and Osborn (2011) correctly point out the need to include hermeneutic phenomenology as a qualitative research methodology to enhance the enactive and embodied research program (see also Gallagher, 2004). And so, in an interesting twist in this work, the truth about saliva, like the truth about the brain, emerges as interpretive, value-laden, and meaningful in relation to the social world.

Note, as in above, the emotional content of the learning situation facilitates a better understanding of the lesson being taught, if you will. As both Heidegger (1926/1996) and Merleau-Ponty (1962/1989) would say, we are meaning-makers in that we unify the relation of our bodies to the world: "Thus experience of one's own body runs counter to the reflective procedure that detaches subject and object from each other, which gives us only the thought about the body, or the body as an idea, and not the experience of the body, or the body in reality" (Merleau-Ponty, pp. 198–199). The disgusting aspect of this praxis provides the necessary tension, born of our bodies, to experience the pre-languaged body in relation to the social and discursive body that *becomes* for others (see Yakhlef, 2010).

In that saliva is not an objective matter, how it matters is meaningful. Saliva is potentially erotic, functional, disgusting, and venomous in a spit; it is drool from a deep sleep or illness, and it clears the mouth of unwanted tastes. Ingold (2001) explains ways of learning as "guided rediscovery" and "To show something to someone is to cause it to be made present for that person, so that he or she can apprehend it directly, whether by looking, listening, or feeling" (p. 141). The saliva demonstration as a pedagogical praxis of "guided discovery" helps students to understand the many ways in which, like saliva, their perceived biological sex type, ethnicity, race, and, in fact, general size and shape are meaningfully shared and constructed through our social interactions and changing beliefs (see also Greene, 1995). Following Merleau-Ponty (1962/1989), we see that what is *natural* about our bodies is made intelligible intersubjectively (i.e., familial, social, and cultural) and is indubitably linked to pre-languaged existence, which is our essential embodiment.

Engaging the learner in active, embodied, and prescribed situated activities is to provide a phenomenological and pedagogical praxis in the classroom setting. First-person qualitative data is then collected from the students and front-loaded into third-person methods (Gallagher, 2003). Thus, we have the phenomenological and the neurological as mutual constraints (Varela, 1996) and mutual enlightenment (Gallagher, 1997) providing teachers with thoughtful and ethical accounts of the relationship between brain, body, and world with regard to adult transformative learning in higher education.

Back to the Brain Itself

What might the adult learner's brain be doing, so to speak, during these praxes? As Juarrero (1999) believes, "From a dynamical perspective, then, learning is the recalibration (in both people and neural nets) of their internal dynamics in response to training" (p. 165). In fact, these learning praxes described above, as part of an ongoing experiential pedagogy, will likely enrich neural connectivity (neural plasticity) in adults through an increase in myelination and in regard to growth-associated proteins and neurogenesis (Gross, 2000; Shaoyu, Chih-hao, Kuei-sen, Guo-li, & Hongjun, 2007; Skene, 1989).

Through excitatory and inhibitory neuro-synaptic processes, neurons assemble and interrelate (LeDoux, 2002); in effect, they collectively generate meaning when in a complex and dynamic interrelationship with the world (Cilliers, 1998). Varela et al. (1991) explain:

> It has, therefore, become increasingly clear to neuroscientists that one needs to study neurons as members of large ensembles that are constantly disappearing and arising through their cooperative interactions and in which every neuron has multiple and changing responses in a context-dependent manner (p. 94).

As Freeman (1999) maintains, "Because brains are composed of interconnected neurons there must be some way in which meanings arise through the activities of neurons" (p. 22). Neural networks may represent the material "prior knowledge" the learner brings to any given learning situation (Zull, 2002). But, we must be careful not to reify the neural network as an unchanging paradigmatic structure. Networks, perhaps, are best interpreted as predispositions that change and evolve in a harmonic interrelation with worldly endeavors.

We remember that Varela et al. (1991) warned against the computer metaphor in cognitive science as perpetuating cognitivism and reductionism, and Davis and Sumara (2000) note that in education this metaphor leads to simple input/output representational theory. Thompson (2007) suggests:

> To describe the brain as a computer in the head whose function is 'information processing' is to reify information into something that preexists 'out there' is 'picked up' and 'processed' by representational systems in the brain, and is independent in principle of the body which serves merely as its 'vehicle.' (p. 186)

If global communication in the brain were restricted to pathways or the "wiring" metaphor, we would not be able to skillfully cope, respond, synthesize, and basically learn with such incredible immediacy. Further, if we adhere to *homunculism*, we would be saddled with the untenable notion that a "little person's" brain inside of us must do some arbitration before *we* can act or think (Descombes, 2001).

Rather, the modus operandi of the higher-order learning and action of the brain is probably facilitated by the phase synchronizing of the frequency of neural oscillations (measured by electroencephalography, EEG). Disparate areas of the brain likely communicate through oscillatory phase locking; the brain has a multiplicity of cadences: waves of rhythms that commune and inform (see Buzsáki, 2006; Varela & Thompson, 2003). Buzsáki (2006) argues that "cortical activity is in perpetual motion and every motor and cognitive act is a synthesis of self-generated, circuit-maintained activity and environmental perturbation" (p. 335). Additionally, Immordino-Yang and Fischer (2011) explain the neuroscience of learning as follows:

> ...learning involves actively constructing neural networks that functionally connect many brain areas. Due to the constructive nature of this process, different learners' networks may differ in accordance with the person's neurological strengths and dispositions, and with the cultural, physical, and social context in which skills are built. (p. 11)

Neurophenomenological Portfolio Assessment

From these praxes described above, teachers as researchers (Kincheloe, 2005; van Manen, 1990) can gather first-person reports of "how" learning took place (an ongoing portfolio) and add these reports to the phenomenological observations carried out by the teacher. A student portfolio assessment can contain student (trained) descriptions of learning experiences, teachers' ethnographic narratives of the learning experience, and third-person corroborative studies (Paulson & Paulson, 1994). This, then, is a phenomenological portfolio assessment based upon these particular praxes and pedagogy. To front-load (Gallagher, 2003) the aforementioned data into an experimental design using third-person methods leads us to a neurophenomenological portfolio assessment.

We can design curricula and assessments based upon phenomenological findings and subsequent evidence from neuroscience. Using these methods, we have attempted to move from the *educable* subject (i.e., the subject of positivism, Cartesianism, and cognitivism) to the *learned agent* understood from a phenomenological and neurological perspective. The value in this is to better understand other ways in which we learn and to expand pedagogy in order to recognize enactive, situated, and embodied forms of learning.

Notes

1. Note that my interest herein is not to reduce learning to categories, types, and styles or to discover learning in the presumed nature and order of the brain. Learning, as I see it, is unique to the person and thusly irreducible and irreplaceable (i.e., no other learns exactly like another). But the aforesaid does not mean learning is simply mysterious, ineffable, and unknowable. Instead, I am interested in expanding notions of learning itself by using neurophenomenological praxis as pedagogy. I choose "praxis" to highlight the hermeneutic aspect that permeates this work. As Bernstein (1983) says, hermeneutics is not "an intellectual stepsister to the methods of natural science" (p. 136). And so, this work makes no claims for advancing an essentialist, objectivist, or positivist account of learning (see also Larkin et al., 2011).
2. This synthetical assessment is highly indebted to the work of Joe Kincheloe (1991, 2005) who has theorized the potential benefits of combining critical constructivism with autopoiesis.
3. In contrast, an objectivist pedagogical assessment of learning categorizes and quantifies the "what" is learned and does so primarily through standardizations of learning (Davis et al., 2005; Kincheloe, 2005).
4. *Educable subject* is the term I will use to describe the subject of power and knowledge within pedagogy (and in some sense in contradistinction to the learning agent). Learning agency then is seen in contrast to the modernist educable Cartesian subject (Davis et al., 2005; Kincheloe, 2005). See also Fendler (1998).
5. For a comprehensive account of learning theories in relation to cognition and education, see Aukrust's (2011) *Learning and Cognition in Education.*
6. Edward Thorndike was a forerunner to American behaviorism. Thorndike examined problem solving by experimenting on animals and extrapolated to humans (Brennan, 1994; Thorndike, 1931; Uttal, 2011). He put animals in experimental conditions that created rewards for specified behaviors. Thorndike noted the measured acquisition of successful responses by trial and error learning. He surmised two basic principles of learning, which he called *exercise* and the *law of effect*. Exercise meant that certain associations were strengthened by repetition and would deteriorate when not used. Thorndike's law of effect stated that when a response was rewarded, it would likely be repeated. However, responses that were punished lessened that particular response. Thus, responses were associated with rewards and punishment. Thorndike, later in his career, adapted the law of effect to show that rewards strengthened associations, but punishment tended to make the learning subject move on to other possible responses. Thorndike's conceptualizing of associative learning left control to the learning subject, whereas Pavlov's (1927/1960) behavioral pairing and conditioning of stimulus and response was in control of the experimenter and so the learning subject (e.g., dogs) would respond and then be given a reward. Thorndike's associative learning, as opposed to Pavlov's reflexology (1927/1960), required the learning subject to be aware of that which reinforced certain responses. Interestingly, Thorndike's learning associations had a cognitive element to them (Brennan, 1994; Uttal, 2011).

7. Pavlov is credited for discovering the essential principles of associative conditioning. He believed that the brain and the nervous system were integral to reflexology. Thus, all learning, for Pavlov, could be reduced to the conjoined relationship between stimuli from the environment and the arbitrating cortex (Brennan, 1994; Uttal, 2011). Tolman's (1922, 1948) understanding of learning was influenced by the behaviorists. However, he offered a gestalt notion of "molar behavior" that described a comprehensive act that was more than the sum of a collection of "molecular" stimuli. For Tolman, "gestalt" described holistic and insightful learning experiences. Tolman is best known for his understanding of learning in terms of the development of cognitive "field" maps, which, presumably, existed in the brain like cognitive schema of the learned environment (Brennan, 1994).
8. For Franz Brentano, psychology was best understood as a science of psychic life as demonstrated in terms of acts (i.e., act psychology). Brentano understood consciousness as a unity recognized by its acts. The psychological act then was intentional in that it pointed toward an *aboutness* within consciousness. The gestalt movement and phenomenological psychology both owe a debt to Brentano (Brennan, 1994).
9. To be learn*ed* is an apt term because it encompasses the noun and action-verb of learning. To be learned means to have had a "history of interactions" in which this "tacit dimension" (Polanyi, 1966) shapes ongoing and future perceptions and thinking (Maturana & Varela, 1987) and provides a "tacit foreknowledge of yet undiscovered things" (Polanyi, 1966, p. 23). This tacit learning dimension is the potential foundation of all further cognitions, and it does not assume, as Piaget's model has, that learning proceeds logically.
10. Today's super computers that model neural networks do learn and are autopoietic (see Winograd & Flores, 1987). But, of course, the computer lacks mood, ethical commitment, value, embodiment, desire, and passion (see Dreyfus, 1972/1992).
11. See also Gibson (1979) on "mutuality."
12. This is why Derrida (1973) says "there never has been any perception" (p. 93).
13. While it is true that Descartes implicated much in his philosophizing of the mind (e.g., the res cogitans beyond res extensa and God), toward the end of his *Sixth Meditation* he looks, briefly, to the brain. Descartes (1637/1980) explains: "...my mind is not immediately affected by all the parts of my body, but merely by the brain...namely, by that part in which the 'common sense' is said to be found" (p. 98).
14. See Husserl (1913/1998) on "sedimentation."
15. William James (1890), John Dewey (1938), and Maurice Merleau-Ponty (1962) all believed we could outline the associations related to biological, neurological, and embodied habit like a path in the woods routinely trodden (see Nöe, 2009; Thompson, 2007). Amazingly ahead of his time, James (1890/2007) explained that "If habits are due to the plasticity of materials [i.e., the brain] to outward agents, we can immediately see to what outward influences, if any, the brain matter is plastic" (p. 107).
16. Neurophenomenology is humanistic and person centered in that it seeks to liberate people from the, more often than not, oppressive strategies of reductionism, objectivism, and determinism sometimes found in neuroscience's accounts of human experience. Phenomenology alone, through its explication of our unique lived experience and shared experiences (as opposed to normalizing and standardizing experiences), is a person-centered methodology and method.
17. Phenomenology is ontological when it is "the science of the being of beings," and it is interpretive and fundamentally hermeneutic when describing is understanding and is always already interpreting (Heidegger, 1926/1996, p. 33). Don Ihde (1986) writes "Thus the *epoché* and *phenomenological reductions* may also be called hermeneutic rules, since they provide the shape or focus of inquiry" (p. 32; see Finlay, 2009).
18. Barclay (2000) explains that "Cognitive neuroscience seems to have appropriated phenomenological insights but ignores some of the philosophical cautions regarding the influence of the 'scientific' perspective when it functions as a presupposition" (p. 142).

19. Wertz (1993) goes on to point out, “Phenomenological psychology does not dismiss the findings and theories of cognitive psychology a priori as untrue or useless. It places them in abeyance while essential insights concerning the psychological sphere are pursued to their limits” (p. 22).
20. I say fruitful, not necessarily equal (see Clegg, 2006). For Clegg, “Both naturalism and phenomenology are foundational ontologies whose conglomeration can result only in the ultimate subjugation of one or the other” (2006, p. 341).
21. For an exhaustive account of the issues regarding Husserlian phenomenology and naturalization, see Roy, Petitot, Pachoud, and Varela (1999).
22. The early Husserl (1900/1973) embarked upon a “logical investigation,” which prescribed a way of putting aside prejudices that would allow the phenomenologist to get closer to the pure phenomenon. I take this beginning version of the epoché to be akin to experimental and disciplinary objectivity (Megill, 1994), where steps are taken to remain neutral to the investigation of a phenomenon. Later Husserl (1913/1998) introduced recognizing and putting aside the natural attitude. Rather than merely putting aside prejudices, the phenomenologist now transcends their entire presumptive framework about reality itself. The phenomenologist then wakes up, if you will, to the transcendental ego that takes part in constituting reality. I suggest transcendence of this sort is a continuum, especially in light of postmodern and post-phenomenological (Ihde, 1986, 1993) assertions (see Finlay, 2009).
23. Phenomenology is not the first-person report itself. One must enact phenomenology before first-person reports are attempted. So where is phenomenology in first-person reports? It is in the phenomenologically informed preparation of the questions that participants are asked; it is in the phenomenological analysis of the responses, which must include the epoché and phenomenological reductions. As van Manen (1990) explains, “From a phenomenological point of view, we are not primarily interested in the subjective experiences of our so-called subjects…the deeper goal, which is always the thrust of phenomenological research, remains oriented to asking the question of what is the nature of this phenomenon…as an essentially human experience” (p. 62). In a phenomenological investigation, first-person reports are not intended to empirically verify a correspondence between subject and object; rather, participants are trained to authentically report the constitutive meaning-making that they contribute to the experience (see Petitmengin & Bitbol, 2009).
24. Gallagher and Zahavi (2008) point out that neurophenomenology began as “an approach to the neuroscience of consciousness” that used phenomenology and later became “any kind of appeal to first-person data in combination with data from neuroscience” (p. 41). One critical issue we encounter is that brain scans are not themselves veridical, apodictic truths. As Fisher (1997) succinctly explains, “The problem is that brain facts are not self-evident. Because no such facts can be found in one’s practical, lived experience, a method is required to reveal them” (p. 49).
25. At this point, it is important to note that phenomenological reflection, the epoché, and phenomenological reductions ought to be the necessary foundation to neurophenomenology. Barclay (2000) questions “if the realm of cognitive science and philosophical cognitivism might have subsumed phenomenological insights into its empirical approach while leaving aside the aspects of phenomenology, which emphasize the reduction of the ‘natural attitude’ by epoché” (p. 162; see also Clegg, 2006).
26. See Cilliers (1998) for a clear and thorough understanding of the relationship between self-organization (autopoiesis) and complex systems.
27. These teaching strategies have been adapted through the years, and, in general, I have picked them up from conversations with colleagues through the years. They are not original to me.
28. Compare these pedagogical observations with “tracking” as described by Siegel (2010).
29. Sabotage is a “naturalistic environment teaching strategy” that is usually done with young children and often with children with certain learning problems and differences (see Ostrosky & Kaiser, 1991).

30. Compare to Hebbian learning and synaptic plasticity (See Freeman, 1999; Hebb, 1949; LeDoux, 2002). Basically, Hebb (1949) suggested strengths and weaknesses of synapses form from use. Changeux and Danchin (1976) further suggested a "use it or lose it" notion regarding an ongoing synaptic pruning.
31. To assess learning is inseparable from the interpretation of learning. Once observed by a third-person (e.g., teacher), learning may be interpreted as knowing, and "knowing is effective action" (Maturana & Varela, 1987, p. 29). Further, the teacher as an interpreter of learning becomes part of the dynamic system of learning, namely, as being a perturbation or attractor within the learning situation (see Juarrero, 1999).

References

Abraham, J. (2003). Dynamic systems theory: Applications to pedagogy. In T. Schacher & A-P., Dauwalder (Eds.), *The dynamical systems approach to cognition: Concepts and empirical paradigms based on self-organization, embodiment and coordination dynamics* (pp. 295–308). Singapore: World Scientific Publishing Company.

Aukrust, V. G. (Ed.). (2011). *Learning and cognition in education.* London, UK: Elsevier Academic Press.

Barclay, M. W. (2000). The inadvertent emergence of a phenomenological perspective in the philosophy of cognitive psychology and psychoanalytic developmental psychology. *Journal of Theoretical and Philosophical Psychology, 20*(2), 140–166.

Bartlett, F. (1932). *Remembering.* Cambridge, MA: Cambridge University Press.

Bateson, G. (1972/2000). *Steps to an ecology of mind: Collected essays in the anthropology, psychiatry, evolution and epistemology.* Chicago, IL: University of Chicago Press

Berger, P. L., & Luckmann, T. (1966). *The social construction of reality: A treatise in the sociology of knowledge.* New York, NY: Anchor Books.

Bernstein, R. J. (1983). *Beyond objectivism and relativism: Science, hermeneutics, and praxis.* Philadelphia, PA: University of Pennsylvania Press.

Blakemore, S. J., & Frith, U. (2005). *The learning brain: Lessons for education.* New York, NY: Wiley-Blackwell.

Bredo, E. (1994). Reconstructing educational psychology: Situated cognition and Deweyian pragmatism. *Educational Psychologist, 29*(1), 23–35.

Brennan, J. F. (1994). *History and systems of psychology* (4th ed.). Englewood Cliffs, NJ: Prentice Hall.

Breunig, M. (2005). Turning experiential education and critical pedagogy theory into praxis. *Journal of Experiential Education, 28*(2), 106–122.

Bruner, J. S. (1968). *Toward a theory of instruction.* New York, NY: W.W. Norton & Co.

Bruner, J. S. (1973). *Beyond the information given: Studies in the psychology of knowing.* New York, NY: W.W. Norton & Co.

Bruner, J. S. (1996). *The culture of education.* Cambridge, MA: Harvard University Press.

Burr, V. (2002). *The person in social psychology.* New York, NY: Taylor & Francis.

Buzsáki, G. (2006). *Rhythms of the brain.* New York, NY: Oxford University Press.

Caine, R. N., & Caine, G. (1991). *Making connections: Teaching and the human brain.* Menlo-Park, CA: Innovative Learning Publications.

Caputo, J. D. (1987). *Radical hermeneutics: Repetition, deconstruction and the hermeneutic project.* Bloomington, IN: Indiana University Press.

Changeux J-P., & Danchin, A. (1976). Selective stabilisation of developing synapses as a mechanism for the specification of neuronal networks. *Nature, 264*, 705–712.

Changeux, J-P., & Ricoeur, P. (2000). *What makes us think? A neuroscientist and a philosopher argue about ethics, human nature, and the brain.* Princeton, NJ: Princeton University Press.

Cho, J., & Trent, A. (2006). Validity in qualitative research. *Qualitative Research, 6*(3), 319–340.

Churchill, S., & Wertz, F. J. (2001). An introduction to phenomenological research in psychology: Historical, conceptual, and methodological foundations. In K. J. Schneider, J. F. T. Bugental, & J. F. Pierson (Eds.), *The handbook of humanistic psychology: Leading edges in theory, research and practice* (pp. 247–262).Thousand Oaks, CA: Sage Publications.
Cilliers, P. (1998). *Complexity and postmodernism: Understanding complex systems*. New York, NY: Routledge.
Clancey, W. J. (2008). The scientific antecedents of situated cognition. In P. Robbins & M. Aydede (Eds.), *The Cambridge handbook of situated cognition* (pp. 11–34). Cambridge, MA: Cambridge University Press.
Clark, A. (2001). *Mindware: An introduction to the philosophy of cognitive science*. New York, NY: Oxford University Press.
Clark, A. (1997). *Being there: Putting brain, body, and the world together again*. Cambridge MA: The MIT Press.
Clegg, J. W. (2006). Phenomenology as foundational to the naturalized consciousness. *Culture and Psychology, 12*(3), 340–351.
Costall, A. (1995). Socializing affordances. *Theory & Psychology, 5*, 467–481.
Costall, A., & Still, A. W. (1991). Introduction: Cognitivism as an approach to cognition. In A. Still, & A. Costall (Eds.), *Against cognitivism: Alternative foundations for cognitive psychology* (pp. 1–6). Hertsfordshire, UK: Harvester Wheatsheaf.
Cox, R. F. A., & Smitsman, A. W. (2008). Towards an embodiment of goals. *Theory & Psychology, 18*(3), 317–340.
Cranton, P. (1996). *Professional development as transformative learning: New perspectives for teachers of adults*. San Francisco, CA: Jossey-Bass.
Daniels, H. (2008). *Vygotsky and research*. New York, NY: Routledge.
Davis, A. B., & Sumara, D. J. (2000). Curriculum forms: On the assumed shapes of knowing and knowledge. *Journal of Curriculum Studies, 32*, 821–845.
Davis, B., Sumara, D., & Luce-Kaplan, R. (2005). *Engaging minds. Learning and teaching in a complex world*. Mahwah, NJ: Lawrence Erlbaum.
Dennett, D. (1991). *Consciousness explained*. Boston, MA: Little, Brown and Co.
Depraz, N. (1999). The phenomenological reduction as praxis. *Journal of Consciousness Studies, 6*(2–3), 95–110.
DeRobertis, E. M. (2011). Existential-humanistic and dynamic systems approaches to child development in mutual encounter. *The Humanistic Psychologist, 39*, 3–23.
Derrida, J. (1973). Difference. In *Speech and phenomena and other essays on Husserl's theory of signs* (D. B. Allison & N. Garver, Trans.). Evanston, IL: Northwestern University Press.
Descartes, R. (1980). *Discourse on method and meditations on first philosophy*. Indianapolis, IN: Hackett Publishing Co. (Original work published 1637/1641)
Descombes, V. (2001). *The mind's provisions: A critique of cognitivism* (S. A. Schwartz, Trans.). Princeton, NJ: Princeton University Press.
Dewey, J. (1938). *Experience and education*. New York, NY: Simon and Schuster.
Dewey, J., & Bentley, A. F. (1949). *Knowing and the known*. Boston, MA: Beacon Press.
Dirkx, J. M. (2002). The power of feelings: Emotion, imagination, and the construction of meaning in adult learning. *New Directions for Adult and Continuing Education, 89*, 63–72.
Dreyfus, H. (1982). Introduction. In H. Dreyfus (Ed.), *Husserl, intentionality and cognitive science*. Cambridge, MA: The MIT Press.
Dreyfus, H. (1992). *What computers can't do: A critique of artificial intelligence*. New York, NY: Harper and Row. (Original work published 1972)
Ellis, R. D. (1999). Integrating neuroscience and phenomenology in the study of consciousness *Journal of Phenomenology, 30*(1), 18–47.
Fendler, L. (1998). What is impossible to think? A genealogy of the educated subject. In T. S. Popkewitz, & M. Brennan (Eds.), *Foucault's challenge: Discourse, knowledge, and power in education* (pp. 39–62). New York, NY: Teachers College Press.
Fenwick, T. (2000). Expanding conceptions of experiential learning: A review of the five contemporary perspectives on cognition. *Adult Education Quarterly, 50*(4), 243–273.

Finlay, L. (2009). Debating phenomenological research methods. *Phenomenology & Practice, 3*(1), 6–25.
Fisher, A. M. (1997). Modern manifestations of materialism: A legacy of the enlightenment discourse. *Journal of Theoretical and Philosophical Psychology, 12*(1), 45–55.
Fodor, J. (1981). *Representations: Philosophical essays on the foundations of cognitive science*. Cambridge, MA: The MIT Press.
Foucault, M. (1980). Two lectures in power and knowledge. In C. Gordon (Ed.), *Power/knowledge: Selected interviews and other writings* (pp. 78–108). New York, NY: Harvester Press.
Freeman, W. J. (1999). *How brains make up their minds*. London, UK: Weidenfeld & Nocolson.
Freeman, W. J. (2000). Emotion is essential to all intentional behaviors. In M. D. Lewis & I. Granic (Eds.), *Emotion, development and self-organization: Dynamic systems approaches to emotional development* (pp. 209–235). New York, NY: Cambridge University Press.
Freire, P. (1971). *Pedagogy of the oppressed*. New York, NY: Seabury Press.
Gadamer, H. (1976). *Philosophical hermeneutics* (D. Lingis, Trans., Ed.). Berkeley, CA: University of California Press.
Gallagher, S. (1997). Mutual enlightenment: Recent phenomenology in cognitive science. *Journal of Consciousness Studies, 4*(3), 195–214.
Gallagher, S. (2003). Phenomenology and experimental design. *Journal of Consciousness Studies, 10*(9–10), 85–99.
Gallagher, S. (2004). Hermeneutics and the cognitive sciences. *Journal of Consciousness Studies, 11*(10–11), 162–174
Gallagher, S. (2005). *How the body shapes the mind*. London, UK; New York, NY: Oxford University Press.
Gallagher, S. (2009). Philosophical antecedents of situated cognition. In P. Robbins, & M. Aydede (Eds.), *The Cambridge handbook of situated cognition*. Cambridge, MA: Cambridge University Press.
Gallagher, S. & Sørensen, J. P. (2006). Experimenting with phenomenology. *Consciousness and Cognition, 15*, 119–134.
Gallagher, S., & Zahavi, D. (2008). *The phenomenological mind: An introduction to philosophy of mind and cognitive science*. New York, NY: Routledge.
Gardner, H. (1983). *Frames of mind and the theory of multiple intelligences*. New York, NY: Basic Books.
Gardner, H. (1991). *The unschooled mind: How children think and how schools should teach*. New York, NY: Basic Books.
Gergen, K. (1995). Social construction and the educational process. In L. P. Steffe, & J. Gale (Eds.), *Constructivism in education*. Hillsdale, NJ: Lawrence Erlbaum Associates.
Gibbs, R. W., & Van Orden, G. C. (2003). Are emotional expressions intentional? A self-organizational approach. *Consciousness and Emotion, 4*, 1–16.
Gibson, J. J. (1979). *The ecological approach to visual perception*. Boston, MA: Houghton Mifflin.
Giorgi, A. (1975). An application of phenomenological method in psychology. In A. Giorgi, C. Fischer, & E. Murray (Eds.), *Duquesne studies in phenomenological psychology: Volume II* (pp. 72–79). Pittsburgh, PA: Duquesne University Press.
Giorgi, A. (1997). The theory, practice, and evaluation of the phenomenological method as a qualitative research procedure. *Journal of Phenomenological Psychology, 28*(2), 235–260.
Giroux, H. A. (2001). *Theory and resistance in education: Towards a pedagogy for the opposition*. Westport, CT: Bergin & Garvey.
Globus, G. G. (1995). *The postmodern brain: Advances in consciousness research*. Philadelphia, PA: John Benjamins Publishing Company.
Goleman, D. (1995). *Emotional intelligence: Why it can matter more than I.Q.* New York, NY: Bantam Books.
Good, J. M. M. (2007). The affordances for social psychology of the ecological approach to social knowing. *Theory & Psychology, 17*(2), 265–295.
Gordon, D. (1988). Tenacious assumptions in western medicine. In M. Lock & D. R. Gordon (Eds.), *Biomedicine examined* (pp. 19–56). New York, NY: Kluwer Academic Publishers.

Greene, M (1995). *Releasing the imagination: Essays on education, the arts, and social change.* San Francisco, CA: Jossey-Bass Publishers.

Greeno, J. G. (1998). The situativity of knowing, learning and research. *American Psychologist, 53*(1), 5–26.

Gross, C. G. (2000). Neurogenesis in the adult brain: Death of a dogma. *Nature Reviews Neuroscience, 1*, 67–73.

Handley, K., Sturdy, A., Fincham, R., & Clark, T. (2006). Within and beyond communities of practice: Making sense of learning through participation, identity and practice. *Journal of Management Studies, 43*(3), 641–653.

Healey, J. M. (2004). *Your child's growing mind: Brain development and learning from birth to adolescence.* New York, NY: Broadway Books.

Hebb, D. O. (1949). *The organization of behaviour.* New York, NY: John Wiley & Sons.

Heidegger, M. (1996). *Being and time* (J. Stambaugh, Trans.). New York, NY: State University of New York Press. (Original work published 1926)

Hill, L. H. (2001). The brain and consciousness: Sources of information for understanding adult learning. *New Directions for Adult and Continuing Education, 89*, 73–82.

hooks, B. (1994). *Teaching to transgress: Education as the practice of freedom.* New York, NY: Routledge.

Hoshmand, L.T., & Martin, J. (1994). Naturalizing the epistemology of psychological research. *Journal of Theoretical and Philosophical Psychology, 14*(2), 171–189.

Husserl, E. (1973). *Logical investigations* (J. N. Finlay, Trans.). New York, NY: Routledge. (Original work published 1900)

Husserl, E. (1998). *Ideas pertaining to a pure phenomenology and to a phenomenological philosophy* (2nd ed.) (Rojcewicz, R. & Schuwer, A., Trans.). Dordrecht, The Netherlands: Kluwer Academic Publishers. (Originally published 1913)

Hutchins, E. (1995). *Cognition in the wild.* Cambridge, MA: The MIT Press.

Ihde, D. (1993). *Postphenomenology: Essays in the postmodern context.* Evanston, IL: Northwestern University.

Ihde, D. (1986). *Experimental phenomenology: An introduction.* New York, NY: State University of New York Press.

Illeris, A. (2008). Transformative learning in the perspective of a comprehensive learning theory. *Journal of Transformative education, 2*(2), 78–89.

Ingold, T. (2001). From transmission of representations to the education of attention. In H. Whitehouse (Ed.), *The debated mind: Evolutionary psychology versus ethnography.* New York, NY: Berg, Oxford International Press.

Immordino-Yang, M. H., & Fischer, K.W. (2011). Neuroscience bases of learning. In V. G. Aukrust (Ed.), *Learning and cognition in education* (pp. 9–15). London, UK: Elsevier Academic Press.

James, W. (2007). *Principles of psychology* (Vol. 1). New York, NY: Cosimo. (Original work published 1890)

Jarvis, P. (2005). *Lifelong learning and the learning society complete trilogy set: Towards a comprehensive theory of human learning* (Vol. 1). New York, NY: Routledge.

Jarvis, P. (2006). *Towards a comprehensive theory of human learning.* New York, NY: Routledge.

Jensen, E. (1998). *Teaching with the brain in mind.* Alexandria, VA: Association for Supervision and Curriculum Development.

Jensen, E. (2008). *Brain-based learning: The new paradigm of teaching.* Thousand Oaks, CA: Corwin Press.

Jordi, R. (2011). Reframing the concept of reflection: Consciousness, experiential learning, and reflective learning practices. *Adult Education Quarterly, 61* (2), 181–197.

Juarrero, A. (1999). *Dynamics in action: Intentional behavior as a complex system.* Cambridge, MA: The MIT Press.

Karmiloff-Smith, A. (1997). *Beyond modularity: A developmental perspective on cognitive science.* Cambridge, MA: The MIT Press.

Kincheloe, J. L. (1991). *Teachers as researchers: Qualitative inquiry as a path to empowerment.* London, UK; New York, NY: The Falmer Press.

Kincheloe, J. L. (1999). *Rethinking intelligence: Confronting psychological assumptions about teaching and learning*. London, UK: Routledge.

Kincheloe, J. L. (2005). *Critical constructivism*. New York, NY: P. Lang.

Kincheloe, J. L. (2008). *Critical pedagogy*. New York, NY: P. Lang.

Koestenbaum, V. (1997). *The phenomenological sense of John Dewey*. Atlantic Highlands, NJ: Humanities Press.

Köhler, W. (1938). *The mentality of apes*. New York, NY: Liveright.

Köhler, W. (1947). *Gestalt psychology*. New York, NY: Mentes.

Larkin, M., Eatough, V., & Osborn, M. (2011). Interpretive phenomenological analysis and embodied, active, situated cognition. *Theory & Psychology*, *21*(3), 318–337.

Lather, P. (1986). Research as praxis. *Harvard Educational Review*, *56*, 257–277.

Lather, P. (1991). *Getting smart: Feminist research and pedagogy with/in the postmodern*. New York, NY: Routledge.

Lave, J. (1988). *Cognition in practice*. Cambridge, MA: Cambridge University Press.

Lave, J., & Wenger, E. (1991). *Situated learning: Legitimate peripheral participation*. New York, NY: Cambridge University Press.

LeDoux, J. (1996). *The emotional brain*. New York, NY: Simon & Shuster.

LeDoux, J. (2002). *Synaptic self: How our brains become who we are*. New York, NY: Penguin Books.

Lemke, J. L. (1997). Cognition, context, and learning: A social semiotic perspective. In D. Kirshner & J. Whitson (Eds.), *Situated cognition: Social, semiotic and psychological perspectives* (pp. 37–57). Mahwah, NJ: Lawrence Erlbaum Associates.

Lewis, M. D., & Todd, R. M. (2005). Getting emotional: A neural perspective on emotion, intention, and consciousness. *Journal of Consciousness Studies*, *12*(8–10), 213–238.

Luhmann, N. (1990). The autopoiesis of social systems. In N. Luhmann (Ed.), *Essays on self-reference*. New York, NY: Columbia University Press.

Luria, A. R. (1976). *Cognitive development: Its cultural and social foundations* (M. Cole, Ed., M. Lopez-Morillas & L. Solotaroff, Trans.). Cambridge, MA: Harvard University Press.

Lutz, A., & Thompson, E. (2003). Neurophenomenology: Integrating subjective experience and brain dynamics in the neuroscience of consciousness. *Journal of Consciousness Studies 10*, 31–52.

MacKeracher, D. (2004). *Making sense of adult learning*. Toronto, ON: University of Toronto Press.

Martin, J., & Sugarman, J. (1997). Societal-psychological constructionism: Societies, selves, traditions and fusions. *Journal of Theoretical and Philosophical Psychology*, *17*(2), 120–136.

Maso, I. (2003). Necessary subjectivity: Exploiting researchers' motives, passions and prejudices in pursuit of answering 'true' questions. In L. Finlay & B. Gough (Eds.), *Reflexivity: A practical guide for researchers in health and social sciences*. Oxford, NY: Blackwell Publishing.

Maturana, H., & Varela, F. (1980). *Autopoiesis and cognition: The realization of the living*. Dordrecht, UK: Reidel. (Original work published 1972)

Maturana, H., & Varela, F. (1987). *The tree of knowledge: The biological roots of human understanding*. Boston, MA: Shambhala Publications.

Megill, A. (1994). Four senses of objectivity. In A. Megill (Ed.), *Rethinking objectivity* (pp. 37–64). Durham, NC; London, UK: Duke University Press.

Merleau-Ponty, M. (1964). *The primacy of perception*. (J. M. Edie, Ed.). Evanston, IL: Northwestern University Press.

Merleau-Ponty, M. (1968). *The visible and the invisible*. (A. Lingis, Trans.). Evanston, IL: Northwestern University Press. (Original work published 1964)

Merleau-Ponty, M. (1983). *The structure of behavior* (A. L. Fisher, Trans.). Pittsburgh, PA: Duquesne University Press. (Original work published 1942)

Merleau-Ponty, M. (1989). *Phenomenology of perception* (C. Smith, Trans.). London, UK: Routledge. (Original work published 1962)

McInerney, R. G. (2010). Oz never did give nothing to the… Scarecrow: Neurophenomenology and critical pedagogy. *Phenomenology and Practice*, *4*(1), 68–87.

Moustakas, C. E. (1994). *Phenomenological research methods.* New York, NY: Sage Publications.
Mezirow, J. (1991). *Transformative dimensions of adult learning.* San Francisco, CA: Jossey-Bass.
Neisser, U. (1976). *Cognition and reality: Principles and implications of cognitive psychology.* New York, NY: W. H. Freeman & Co.
Noë, A. (2004). *Action in perception.* Cambridge, MA: The MIT Press.
Noë A. (2009). *Out of our heads: Why you are not your brain, and other lessons from the biology of consciousness.* New York, NY: Hill & Wang.
Notterman, J. M. (2000). Note on reductionism in cognitive psychology: Reification of cognitive processes in to the mind, mind-brain equivalence and brain-computer analogy. *Journal of Theoretical and Philosophical Psychology, 20,* 116–121.
Osbeck, L. M. (2009). Transformations in cognitive science: Implications and issues posed. *Journal of Theoretical and Philosophical Psychology, 29*(1), 16–33.
Ostrosky, M. M., & Kaiser, A. P. (1991). Preschool classroom environments that promote communication. *Teaching Exceptional Children,* 2(34), 6–10.
Paré, D., Collins, D. R., & Pelletier, J. G. (2002), Amygdala oscillations and the consolidation of emotional memories. *Trends in Cognitive Sciences, 6,* 306–14.
Paulson, L., & Paulson, P. (1994). Assessing portfolios using the constructivist paradigm. In R. Fogarty (Ed.), *Student portfolios.* Palatine, IL: IRI Skylight Training & Publishing.
Pavlov, I. (1960). *Conditioned reflexes: An introduction of the physiological activity of the cerebral cortex* (G.V. Anrep, Ed., and Trans.). New York, NY: Dover. (Original work published 1927).
Pea, R. (1993). Practices of distributed intelligence and designs for education. In G. Salomon (Ed.), *Distributed cognitions: Psychological and educational considerations.* Cambridge, MA: Cambridge University Press.
Petitmengin, C., & Bitbol, M. (2009). The validity of first-person descriptions as authenticity and coherence. In C. Petitmengin (Ed.), *Ten years of viewing from within: The legacy of Francisco Varela* (pp. 363–404). London, UK: Imprint Academic.
Piaget, J. (1975). *The development of thought: Equilibrium of cognitive structures.* New York, NY: Viking Press.
Pinar, W. F. (2004). *What is curriculum theory?* Mahwah, NJ: Lawrence, Erlbaum.
Polanyi, M. (1966). *The tacit dimension.* New York, NY: Doubleday.
Poerksen, B. (2004). *The certainty of uncertainty: Dialogues introducing constructivism* (A. R. Koeck, & W. F. Koeck, Trans.). London, UK: Imprint Academic.
Rietveld, E. (2008). The skillful body as concernful system of possible actions: Phenomena and neurodynamics. *Theory & Psychology, 18*(3), 341–364.
Robbins, B. D. (2006). An empirical, phenomenological study: Being joyful. In C. T. Fischer (Ed.), *Qualitative research methods for psychologists: Introduction through empirical studies* (pp. 173–212). Burlington, MA: Elsevier, Academic Press.
Rose, S. (2005). *The future of the brain: The promise and perils of tomorrow's neuroscience.* New York, NY: Oxford University Press.
Roy, M-J, Petitot, J., Pachoud, B., & Varela, F. J. (1999). Beyond the gap: An introduction to naturalizing phenomenology. In J. Petitot, F. J. Varela, B. Pachoud & J-M, Roy (Eds.), *Naturalizing phenomenology: Issues in contemporary phenomenology and cognitive science* (pp. 1–82). Stanford, CA: Stanford University Press.
Rudrauf, D., Lutz, A., Cosmelli, D., Lachaux, J-P., & Le Van Quyen, M. (2003). From autopoiesis to neurophenomenology: Francisco Varela's exploration of the biophysics of being. *Biological Research, 36,* 21–59.
Sacks, P. (1999). *Standardized minds: The high price of America's testing culture and what we can do to change it.* Cambridge, MA: Perseus Books.
Sawyer, R. K., & Greeno, J. G. (2009). Situativity and learning. In P. Robbins & M. Aydede (Eds.), *The Cambridge handbook of situated cognition.* Cambridge, MA: Cambridge University Press.
Shaoyu, G., Chih-hao, Y., Kuei-sen, H., Guo-li, M., & Hongjun, S. (2007). A critical period for enhanced synaptic plasticity in newly generated neurons of the adult brain. *Neuron, 54*(4), 559–56.

Sheets-Johnstone, M. (1990). *The roots of thinking*. Philadelphia, PA: Temple University Press.
Siegel, D. (1999). *The developing mind: How relationships and the brain interact to shape who we are*. New York, NY: Guilford Press.
Siegel, D. J. (2010). *The mindful therapist: A clinician's guide to mindsight and neural integration*. New York, NY: W.W. Norton and Co.
Simms, E. M. (1999). The countryside of childhood: A hermeneutic phenomenological approach to developmental psychology. *The Humanistic Psychologist, 27*(3), 301–328.
Skene, J. H. (1989). Axonal growth-associated proteins. *Annual Review of Neuroscience, 12*, 127–156.
Skinner, B. F. (1953). *Science and human behavior*. New York, NY: Macmillan Press.
Skinner, B. F. (1974). *About behaviorism*. New York, NY: Alfred A. Knopf
Sousa, D. A. (2010). *Mind, brain and education. Neuroscience implications for the classroom*. Bloomington, IN: Solution Tree Press.
Spinelli, E. (1989). *The interpreted world: An introduction to phenomenological psychology*. London, UK: Sage.
Szasz, T. (1996). *The meaning of mind: Language, morality, and neuroscience*. Westport, CT: Praeger Publishing.
Thelen, E., & Smith, L. (1994). *A dynamic systems approach to the development of cognition and action*. Cambridge, MA: The MIT Press.
Thibault, P. J. (2004). *Brain, mind, and the signifying body: An ecosocial semiotic theory*. New York, NY: Continuum Publishing.
Thompson, E. (2007). *Mind in life: Biology, phenomenology, and the sciences of the mind*. Cambridge, MA: Harvard University Press.
Thompson, E., Lutz, A., & Cosmelli, D. (2005). Neurophenomenology: An introduction for neurophilosophers. In A. Brook & K. Akins (Eds.), *Cognition and the brain: The philosophy and neuroscience movement* (pp. 40–97). New York, NY: Cambridge University Press.
Thorndike, E. L. (1931). *Human learning*. New York, NY: Appleton.
Tolman, E. C. (1922). A new formula for behaviorism. *Psychological Review, 29*, 44–53.
Tolman, E. C. (1948). Cognitive maps in rats and men. *Psychological Review, 55*, 189–208.
Toren, C. (2001). The child in mind. In H. Whitehouse (Ed.), *The debated mind: Evolutionary psychology versus ethnography* (pp. 155–179). New York, NY: Berg, Oxford International Press.
Uttal, W. R. (2011). *Mind and brain: A critical appraisal of cognitive neuroscience*. Cambridge, MA: The MIT Press.
Vanderstraeten, R. (2002). Dewey's transactional constructivism. *Journal of Philosophy of Education, 36*(2), 233–246.
van Gelder, T. J. (1998). The dynamical hypothesis in cognitive science. *Behavioral and Brain Sciences, 21*, 1–14.
van Manen, M. (1990). *Researching lived experience: Human science for an action sensitive pedagogy*. New York, NY: State University of New York Press.
Varela, F. J. (1996). Neurophenomenology: A methodological remedy for the hard problem. *Journal of Consciousness Studies, 3*, 330–350.
Varela, F. J. (1999). The specious present: A neurophenomenology of time consciousness. In J. Petitot, F. J. Varela, B. Pachoud, & J-M. Roy (Eds.), *Naturalizing phenomenology: Issues in contemporary phenomenology and cognitive science* (pp. 266–314). Stanford, CA: Stanford University Press.
Varela, F. J., & Shear, J. (1999a). First-person methodologies: What, why, how? *Journal of Consciousness Studies, 6*(2–3), 1–14.
Varela F. J., Lachaux J. P., Rodriguez, E., & Martinerie, J. (2001). The brainweb: Phase synchronization and large-scale integration. *National Review of Neuroscience, 2*(4), 229–39.
Varela, F. J., & Thompson, E. (2003). Neural synchrony and the unity of mind: A neurophenomenological perspective. In A. Cleeremans (Ed.), *The unity of consciousness: Binding, integration and dissociation* (pp. 266–287). New York NY: Oxford University Press.

Varela, F. J., Thompson, E., & Rosch, E. (1991). *The embodied mind: Cognitive science and human experience*. Cambridge, MA: The MIT Press.
Vygotsky, L. (1978). *Mind in society: The development of higher psychological processes*. Cambridge, MA: Harvard University Press.
Waldrop, M. M. (1992). *Complexity: The emerging science at the edge of order and chaos*. New York, NY: Simon & Schuster.
Walkerdine, V. (1984). Developmental psychology and the child-centered pedagogy: The insertion of Piaget into early education. In J. Henriques, W. Hollway, C. Urwin, C. Venn, & V. Walkerdine (Eds.), *Changing the subject: Psychology, social regulation and subjectivity* (pp. 153–202). UK, Methuen; New York, NY: Routledge.
Walkerdine, V. (1997). Redefining the subject in situated cognition theory. In D. Kirshner & J. Whitson (Eds.), *Situated cognition: Social, semiotic and psychological perspectives* (pp. 57–70). Mahwah, NJ: Lawrence Erlbaum Associates.
Wertz, F. (1993). Cognitive psychology: A phenomenological critique. *Journal of Theoretical & Philosophical Psychology*, *13*(1), 1–24.
Wilson, E. A. (1998). *Neural geographies*. London, UK: Routledge.
Wink, J. (2010). *Critical pedagogy: Notes from the real world*. New York, NY: Prentice Hall.
Winograd, T., & Flores, F. (1987). *Understanding computers and cognition: A new foundation for design*. Indianapolis, IN: Addison-Wesley Professional
Yakhlef, A. (2010). The corporeality of practice-based learning. *Organization Studies*, *31* (4), 409–430.
Zahavi, D. (2003). *Husserl's phenomenology*. Stanford, NY: Stanford University Press.
Zull, J. E. (2002). *The art of changing the brain: Enriching teaching by exploring the biology of learning*. Sterling, VA: Stylus Publishing.

Cognitive Phenomenology in the Study of Tibetan Meditation: Phenomenological Descriptions Versus Meditation Styles

Olga Louchakova-Schwartz

Phenomenology… offers detailed analyses of various aspects of consciousness, including perception, imagination, embodiment, memory, self-experience, temporality, etc. In offering such analyses, phenomenology addresses issues that are crucial for an understanding of the true complexity of consciousness and might even offer a conceptual framework for understanding the mind that is of considerably more value than some of the models currently in vogue in cognitive science.

(Zahavi, 2010, p. 8)

In spite of neurophenomenology's promise to answer the burning questions of the study of consciousness, the use of neurophenomenology in experimental science remains limited (Overgaard, 2004). After decades of behaviorism, researchers tend to dismiss first-person research methods.[1] Arguing against such perspective, Thompson says: "Neurophenomenology stresses the importance of gathering first-person data from phenomenologically trained subjects as a heuristic strategy for describing and quantifying the physiological processes relevant to consciousness" (Lutz & Thompson, 2003, p. 32). With the recent rapid growth of interest in meditation, first-person reporting appears especially necessary (Lutz, Dunne, & Davidson, 2007). In this chapter, I argue that first-person reporting and subsequent phenomenological analysis are necessary for the experimental study of meditation. As an example, I will use our recent study of visual-spatial processing in Tibetan monks (Kozhevnikov, Louchakova, Josipovic, & Motes, 2009). The phenomenological part of this research will be reported here for the first time.

O. Louchakova-Schwartz, M.D., Ph.D. (✉)
Sofia University (Formerly Institute of Transpersonal Psychology),
1069 East Meadow Circle, Palo Alto, CA 94303, USA
e-mail: olouchakova@gmail.com

S. Gordon (Ed.), *Neurophenomenology and Its Applications to Psychology*,
DOI 10.1007/978-1-4614-7239-1_3,

There is a general scientific belief that subjective experience has little to do with cognition.[2] To explain the mind, cognitive science uses the computer metaphor, according to which cognition consists of information processing, which only appears as subjective experience. Thereby, research should target the cognitive processing, not the subjective experience. However, new brain-imaging technologies demonstrate that there are connections between the activities of the brain and subjectively experienced mental states. Experience appears to have a robustness (repetitive and reproducible quality) that is intimately related to brain functioning. Experience also has an important influence on behavior,[3] and the latter is something we can measure, assess, and evaluate by various tests and stimulus–response experiments. Consequently, there is the possibility of establishing a connection between the purely subjective life of consciousness and objective behavior through neurophenomenological research. The main difficulty in designing such research consists of articulation and averaging the experience. In meditation research, both problems are less prominent due to the disciplined character of introspection. Teachers of meditation frequently have the skills of verbalizing internal experience. Difficulties in verbalization can be bypassed by the use of negation, simile, and metaphor.[4]

The juxtaposition of the concrete and logically coherent environment of the scientific experiment with the messy contents of experience causes the mutual clarification of experience and neurocognitive theory.[5] Gallagher (1997) calls this a "mutual enlightenment" (p. 195). In the research of meditation, the phenomenology of meditation experience serves in place of Occam's razor, helping to reify interpretations, theories, and hypotheses.[6]

In the last couple of years, it was shown that the research of meditation contributes to scientific studies of attention, consciousness, self-awareness, empathy, and the formation of the theory of mind (Cahn & Polich, 2006; Green & Turner, 2010; Newberg, 2010; Raffone & Srinivasan, 2010). On one hand, meditation is a natural and ordinary experience, and on the other hand, not everybody is familiar with the states of experience which arise in meditation.[7] As opposed to the prevailing opinion that meditation silences the mind, subjective experiences in meditation can be very complex. Firstly, meditation is a dynamic phenomenon: the experience of meditation can take different forms depending on the type of practice, character, psychological maturity, and proficiency of the meditator. The experience of meditation changes over time and also within a single session. Secondly, meditative experience creates a new spectrum of cognitive possibilities, e.g., experiences of heightened intuition or insight. Thirdly, the experience of meditation brings with it self-knowledge which is different from self-knowledge in the natural attitude. Self-knowledge in meditation is based on the direct intuition of the contents of consciousness, while in the natural attitude, self-knowledge is the knowledge of one's behavior. Since personal experience informs our understanding of consciousness, meditators and non-meditators have different theories of mind and even different personal ontologies.[8] There is a definitive "experience gap" between meditators and non-meditators, and it can be reflected in their views on cognition.

For example, Buddhist scholars, including the Dalai Lama and Mathieu Ricard, claim that the advanced practitioners can hold in their minds a detailed complex image of a mandala, a symbolic depiction of the universe inhabited by different beings, for many minutes or even hours without a noticeable degradation (Barinaga, 2003; Kozhevnikov et al., 2009). However, at the time of the dispute between scientists and Buddhist scholars in 2003 at MIT, the scientific theories of visual cognition indicated that this should not be possible. Visual imagery (the images participating in visual perception) and internal mental imagery are common to a majority of people. Ordinarily, images change in a matter of seconds. The same brain networks run both mental and visual imagery, and these networks are set up for the fast fading of images (Kingsbourne, 2006; Kosslyn, 1995). As a means of engagement with the outside world, vision would not be possible otherwise.

When cognitive theories of mental imagery were published in the mid-1990s, meditation was thought of as a mere relaxation technique.[9] The internal experiences of meditators were outside the scope of attention in science. Consequently, scientific views on mental imagery were informed by the common experience of the rapidly fading images (cf. Kosslyn, Seger, Pani, & Hillger, 1990) and not by the experience of the stable images in visual meditation. The majority of researchers agree that the internal mental imagery originates from the memories of prior experiences. The complexity and duration of the internal image depends on the functioning of the visual working memory (VWM). Like the virtual memory in a computer, VWM maintains operational information about the stimuli no longer in view. The complexity and duration of an image depends on how much information the VWM can hold. The mind is strikingly limited in that regard: VWM is a part of overall working memory, and the capacity of working memory is consistently estimated between two and four items. Working memory processes the spatial and visual (appearance-related) aspects of an image separately. It is believed that spatial memory can be increased by the repeated practice, e.g., via videogames, but this will not affect the memory related to the appearance of the image.[10] In order to explain the claims to extraordinary imagery capacity in Tibetan monks, the present researchers (Kozhevnikov et al., 2009) decided to test cognitively whether the Buddhist meditation training could enhance the VWM with regard to both the dynamic manipulation of spatial information and the maintenance of a simple image.

An intention to research the visual imagery in meditation met with the following difficulty. Tibetan Buddhism includes hundreds of meditation techniques, which can be combined and used for many different purposes; not all of the techniques cause the reported enhancement of mental imagery. Tibetan meditation does not consist of a simple set of repetitions, is very different from videogames, and does not use pharmaceuticals, the traditional devices in VWM experiments. Besides, training in meditation is individualized according to the capabilities, character, mental qualities, and advancement of the particular practitioner. Therefore, an effective research design that should be able to resolve the conflict between theory and experience needed the information about what kind of experience should be subjected to cognitive testing. Can this experience be articulated, and will it be possible to find the cognitive correlates to such experience?[11] It was necessary first to clarify what kind of Tibetan meditation causes an empirical enhancement of mental imagery and then find the ways to link the data of experience to cognitive testing. To show the logistics of developing a combined design that included phenomenological analysis and cognitive testing, I will first have to briefly review how the meditation experience was accounted for in the prior research.

Experience of Meditation and Scientific Research

In meditation the mind is aware of its own internal processes. It consciously monitors and regulates itself through self-generated influences (Wittig, 2001, p. 97). Similar to dreaming, meditation is a purely subjective state, with the difference that in meditation one is consciously self-monitoring. In meditation, the initially blank space of introspection is over time more and more animated with images, thoughts, and feelings (Louchakova, 2007b, 2007c). Meditation enriches the contents of the mind by bringing up new elements in internal experience, such as the awareness of the internal space, mental images, or new memories. It can also bring up previously unnoticed structural aspects of experience such as differentiation between the awareness and its objects, or the preconceptual state of pure subjectivity.[12] Different spiritual meditative traditions regard this state as a necessary component of spiritual completion and happiness and associate it with enlightenment, salvation, liberation, and self-knowledge. These meaningful contexts contribute to the meditation process: research has showed that meditation with spiritual goals produces stronger positive physiological and psychological effects.[13]

The intentional modification of internal experience in meditation can shift, expand, or focus attention; relax the mind; and introduce the attentional modes of witnessing, receptivity, and/or letting go. Since these activities are internal, without external cognitive or behavioral output, meditation was linked to the so-called intrinsic brain state, the conscious activity of the mind outside of the stimulus–response situation.[14] While the mind is not engaged in external tasks, meditation involves the "inner doing" that creates the meditation experience. Specific experience is simultaneously an existential goal of meditation, an evidence of its deeper cognitive effects, and an active agent of transformation. In addition to this complexity, the whole internal landscape, the proficiency in self-regulation, and the degree of presence to the internal processes change as the practice progresses over time.[15]

In early research, the content of internal experience was not viewed as important, and meditation was treated as a simple relaxation technique.[16] Soon it became obvious that the effects of meditation on the mind are at least as important as its effects on the body.[17] Meditation was reinterpreted as a cognitive and attentional strategy that enhances perception, emotion, cognition, and learning via optimization of the executive attention network and top-down regulation of physiological processes.[18] However, the studies supporting these interpretations remain internally noncohesive due to the diversity of research methods and methodologies and disparity of findings.[19]

Disparity of findings would not be a problem for the study of conscious experience in general, but for a relatively unknown terrain of experience, such as meditation, the situation is different. Cognitive theories view the mind as a set of information-processing modules. The computer metaphor can accommodate very diverse theoretical and mathematical processing models for this modular functioning. The models are constructed on different principles and can explain any given data set.[20] Since the research designs and research testing tools are themselves based on cognitive theories, the situation is a bit like a dog chasing its tail: the theory produces the findings, and the findings feed into the theory that produced them. For meditation research, this creates a situation which limits the possibility of the new and critical findings.

For example, an EEG researcher of meditation may be a localizationist who believes that the brain functions as a set of centers or a connectionist who believes that the brain functions as a network.[21] In the localizationist paradigm, the researcher will measure the event-related potentials and test the separate attentional networks. In the connectionist paradigm, he or she may analyze the nonlinear dynamics of EEG or coherences in the neural networks. Different analytic paradigms can be applied to the same EEG recording and render completely different data sets, internally congruent, but not connected with one another. Moreover, there may be changes in one paradigm and no changes in the other. Which one gives a more adequate picture of what's going on? Although the heuristics of the research process always updates the theories, in general the cognitive scientific research tends to be theory-laden. A theoretical framework in cognitive neuroscience serves as a filter that both structures research and interprets the findings in relative isolation from another framework.

The knowledge of experience in meditation helps both to choose a correct research paradigm for the study of meditation and to refine cognitive theories by pointing out to the non-ordinary aspects of the brain and expanded possibilities of cognition. Large and still growing evidence shows that the specific experiences of consciousness in meditation are steadily accompanied both by the particular changes in EEG spectra and by the specific changes in the indices of electrical activity such as alpha-blocking or connectivity of the default networks.[22] Recently, a high correlation between the internal shifts of consciousness in meditation and the metabolic activity of different brain areas was demonstrated through fMRI (Baerentsen et al., 2010). Additionally, it was shown that besides correlations with the neuroactivity, subjective experience in meditation is linked to neuroplasticity and neurogenesis in the areas of the brain involved in meditation.[23]

Among many ways to study experience, meditation experience was approached mostly in the natural attitude, through anecdotal evidence and ethnographic observations.[24] Phenomenological, i.e., strict systematic reflective, analysis of the constitutive structures of experience in meditation took place in only a few studies.[25] Since we know little about the phenomenology of meditation, scientific reports have indicated at the most whether meditation comes from Yoga, Zen, or other tradition, leaving out what the meditators actually did with their minds and what kind of experience they attained. While researchers cannot follow these changes within the stimulus–response cognitive research design,[26] brain imaging is catching up with empirical observations. Sometimes it takes many years to discover a connection between an experientially available internal state and its brain correspondent. For example, Travis and Wallace's (1999) account of the experientially perceived typical shift at the entry into a meditative state was published 10 years prior to Baerentsen et al. (2010), who discovered the massive corresponding changes in metabolism in specific areas of the brain.

Meditation Styles

Averaging of the data of different kinds of meditation has been a long-standing common problem in meditation research (Travis & Shear, 2010). To solve this problem, attempts were made to develop a classification of meditation suitable for research purposes. The first classification divided meditation into "concentration" and "awareness" or "mindfulness" types (Davidson & Goleman, 1977; Davidson, Goleman, & Schwartz, 1976; Goleman, 1988). This division appears to have been borrowed from Tibetan Buddhism, which differentiates between these two groups of practices. The outcomes of this empirical division will be different in Buddhism and in science. In science, attention and awareness are two close functions, if not the same function. This division indicates that attention/awareness is the key constituent of meditation practice, and therefore that research should focus on the cognitive function of attention. Buddhist metaphysics differently maintains that awareness is not attention and not a cognitive mode, but rather the universal ontological ground. In order to reach this ontological ground, the meditator has to abandon the natural attitude in which the self and the world are two separate realities, and the world consists of material entities. Adopting an ontological focus on experience akin to phenomenological investigations,[27] the meditator's practice brings out of anonymity the essential structures of consciousness and makes the concept of pure awareness empirical reality.[28] These essential structures of experience in meditation include the subject-object differentiation, restriction of attention to the pure subjectivity, focus of attention specifically on one object at the exclusion of other objects, or the presence of a space-like field of awareness in which its referents (objects of experience) appear, etc. Division of meditation into two meditation styles does not accommodate the diversity of rising states and structures of consciousness, some of which are stable and cross over different kinds of meditation and some of which are fluid. How this situation will be translated into the language of cognitive science remains unclear, as the findings from each group of meditation styles reveal much diversity (Neumann & Karel, 2006).

Some researchers attempted classifications that would be closer to the actual experience of meditation. Phenomenologically, meditators transition from the natural state of "I experience this" to a state of witnessing such as "I am aware of this," for example, from "I am angry" to "I am aware of my anger." Although not a fully focused state itself, the witnessing state trains the focusing of the mind. As the practice deepens, focused witnessing shifts into an open, unattached awareness in which all the objects are presented within the expanded field of one's presence (Wangual, 1993). Taking this into account, Lutz suggested further refinement of the categorization of meditation into concentration and open awareness styles (Lutz, Brefczynski-Lewis, Johnstone, & Davidson, 2008). This division allows introducing further distinctions dependent on what objects concentration is devoted to. The focus of the mind on a single object to the exclusion of all other objects may be on sounds, colors, internal lights, elements of the body schema and image, art objects, objects of nature, etc. Depending on the object, the effects of focusing will be different.[29] Even the state of pure open awareness changes over time. Within that state, there are conditions when open awareness includes the objects, and when it does not, or when the meditator's identity shifts to that of pure awareness or completely disappears.[30] When the states of pure awareness emerge in meditation, awareness is both the subject and the object of meditation, and the states shift from dual (with subject-object divisions) to non-dual (Josipovic, 2010).

Given this complexity, several researchers have suggested that instead of maintaining strictly defined categories, such as concentration, mindfulness, or open presence, meditation should be considered on a spectrum of consciousness (Cahn, Delorme, & Polish, 2010). Travis and Shear (2010) believe that this spectrum constitutes the third meditation style, automatic self-transcending, which includes techniques designed to transcend their own activity and therefore to change the experience. In some traditional schools of meditation, such as Buddhist Nyingma, this self-transcending hermeneutics of inner experience is brought to a conceivable maximum; in other schools, such as Indian Vedanta, the process of meditation is much more formalized. Despite these efforts, what is meant by the experience of meditation itself, and what approach would lead to the understanding of this experience, needs to be further clarified. As noted by Guenther, the prominent scholar of Tibetan Buddhism, "The concept of meditation is not precise and eventually will have to be replaced by [another concept, which is] more fully descriptive and elucidative of the processes involved" (1992, p. xi). This highlights the need to obtain first-person data for designing meditation research protocols. As I will show in this study, it is the concrete analysis of the particular meditative experience that proves to be a key component in a successful experimental design.

Toward the Cognitive Phenomenology of Meditation

To be useful to scientific research, the analysis of experience has to be able to reach not only the traditional phenomenological first-person but also the second-person consciousness. This depends on articulation and understanding of the experience of the other on his or her own grounds and therefore needs a framework that helps one understand those experiences, which are other than one's own. Such conceptual clarification and systematic description of experience (Marbach, 2006) is the purview of phenomenology. The history of psychology is filled with examples in which experiences not fitting within the natural attitude were reduced to pathology or neglected. Categories of non-ordinary or exceptional experiences were invented to account for the normalcy of some classes of non-common experiences. One class of experiences, the direct intuition of the essential structures of consciousness, was not addressed in psychology. Normal in the sense that they reveal the essential structures of consciousness, these experiences are not common. However, they are shared between meditation and phenomenology.

Meditative traditions such as Yoga, Vedanta, Vipashyana, or Tibetan or Hindu Tantra train the capacity of systematic apperception of the structures of individual, subjective consciousness toward apprehension of consciousness in pure subjectivity, or consciousness as such. These conditions are devoid of the usual meaningful connotations. They can be described in terms of their existence, sentiency, or fullness, but not in terms of their meaning or in terms of their exact appearances.[31] A parallel tendency in phenomenology also uncovered the pure subjectivity of the ego, and consciousness as such.[32] The similarities between the philosophical phenomenological consciousness and consciousness in meditation were acknowledged by many and do not have to be repeated here. On the contrary, I will mention the differences. Firstly, pure consciousness in meditation is not derived philosophically, but is lived experientially. Secondly, while pure consciousness in a philosophical sense is devoid of qualities, the lived gestalt of such consciousness in meditation is characterized by its great intensity (Guenther, 1992). When all dimensions of time and space are transcended, the meditator experiences her very being as the essence of conscious experiencing in the absence of individuality. After a while, the meditator returns to normal subject-object dichotomization of perception and thought. In this manner, advanced states of meditation fluctuate between the two experientially given modes: one of them an awareness of the experiencer being at the same time the object of experiencing, and the other constituted only of the pure essence of experiencing without any object and agency. In the second state, the experiencer-experiencing-experienced distinction is absent. The absence of intentionality, or "aboutness" of consciousness, is the most persistent feature of the advanced meditative experience. Research of the spectrum of such states requires adaptation of both psychological and cognitivist frameworks. Allegedly, the subjectivity of consciousness is beyond the grasp of objective science (Nagel, 1971). Even if the pure subjectivity of consciousness cannot be reduced to a set of cognitive variables, the "great intensity" of its empirical character should have neurocognitive correlates. According to non-reductive physicalism, different levels of interpretation of experience should be considered as not reducible to one another, but as abiding in correspondence.[33] In neuroscience, there are similar proposals.[34]

Lived experience, philosophical conceptualization of experience, and scientific models come together in cognitive phenomenology. Cognitive phenomenology accounts for the internal states occurring in the absence of behavioral manifestations, encouraging integration of the phenomenological analysis with experimental designs.

Phenomenological-Cognitive Mapping

Tibetan Buddhist training begins with contemplation of positive attitudes and values and proceeds toward deep modifications of the mind. The broad spectrum of practices includes meditations as diverse as witnessing the contents of the mind, gazing into the sky and imagining the sky space extending into one's internal space, processing subtle energies through the system of embodied centers and meridians, and working with the experientially available elements of the self-body-world construct, including movement, light, sound, breath, color, archetypal images, and emotions. Hundreds of meditation practices can literally work off every modality of experience, turning them into the substratum for meditation. The training accounts for individual psychological features and adapts the practice so that it continuously furthers developmental changes. Intense hermeneutics of meditative training (Table 1) distinguishes this tradition from the more formalistic approaches to meditation in Yoga, Sufism, Hesychasm, or Vedanta. The meditation process, experience, technique, and environmental influences are involved in the interactive cycle and are systemically interconnected.

In order to find the training (1), which would be best associated with the effects on visual memory reported by the Dalai Lama, researchers interviewed advanced practitioners of meditation and conducted the phenomenological analysis of their

Table 1 Hermeneutic cycle in Tibetan meditation

(1) Meditation process	(2) Particular meditation technique
Tibetan meditation has a goal of experiential reformulation of all cognitions so that they can be seen as having their origin in pure awareness. (1) It happens over time and is influenced by (2), (3), and (4).	In this approach, the meditator can choose from a spectrum of possibilities of what to do internally with their mind, using various meditation techniques. They can switch techniques over time, or sometimes even in one session, dependent on (1), (3), and (4). Meditation technique is intentional; it influences (1), (2), and (3) and is in turn influenced by each of them.
(4) Meditation experience	(3) Environmental influences
These are contents and dynamic of ones perception of their inner world, various structures of consciousness, events of inner life, and emotional contents. Some of it changes from session to session, some changes over a longer period of time, and some remains the same. Meditation experience is influenced by (2) and (3) and by culminative changes in (1).	Individual fluctuations of the mind Proficiency of meditator/degree of advancement Constitutional differences, such as gender, culture, etc. Some influences are stable, and some are affected by (4), (3), and (1).

Note. Demonstrates the relationship between the goal of the meditation process (1), meditation technique (2), environmental influences (3), and the subjective experience of the meditator (4). These parameters are not fixed, but interrelated and dynamic accounts. Insofar as it is possible, (2), (3), and (4) were taken into consideration in interviewing. Three out of four researchers on the team were advanced meditators themselves (the present author being a certified teacher of Tantric and Vedantic meditation). The present researcher also consulted the classical Buddhological works of Stcherbatsky and Guenther.[35] Research participants (a) were self-recognized as experienced meditators, (b) carried authority granted by their communities, and (c) were recognized as successful meditators by their teachers. All participants have had meditation retreats lasting at least 1 year and indicated that they had more than 10 years of experience in meditation. Their accounts were treated by the genetic phenomenological analysis, with some reference to Giorgi's adaptation of Husserl's phenomenological method for psychology, and Merleau-Ponty's reflective analysis of perception.[36]

Phenomenology of Consciousness in Tibetan Meditation

Tibetan Buddhism distinguishes between the "referential" and "non-referential" modes of consciousness. The referential mode includes all states of consciousness in which awareness has a vector toward the objects.[37] In the non-referential mode, this vector is absent. This is reflected in the meditation practice. Tenzin Wangyal describes the relationship between awareness and its objects in the developing practice as the "sunbeams [awareness] melt the snow [objects]" (referential mode) or the "snowflakes [objects] falling into the ocean [of awareness]" (non-referential mode). Because the separateness of a referent eventually disappears in the experience of the pure objectless awareness, the latter is metaphorically referred to as a lion, or the sun (Guenther, 1992; Wangual, 1993). As a result of practice, self-awareness is reconstituted and the personal transformation takes place.[38]

Tibetan Buddhism has two traditions of meditation training: the older original Tibetan rDzogs-chen and the newer Mahāmudrā with some Indian influences. From the very beginning of the training, rDzogs-chen orients the mind toward the recognition of the perceptual-cognitive state of open, pure awareness devoid of objects (i.e., non-referential mode) and reinterprets all mental processes against this background. By contrast, Mahāmudrā is more preoccupied with the object (Guenther, 1992, p. 39) and features a gradual transition from the referential to the non-referential mode. Our group of participants included practitioners from both traditions.

Meditative concentration of the mind on a mental image takes place in the respective contexts of both systems. In Mahāmudrā, concentration plays the leading role, whereas in rDzogs-chen, this is necessary only insofar as the mind gains strength and unity necessary for the gestalt of pure awareness (Rig-pa). The essence of the process is not the focusing of attention, but rather a decentering of strengthened attention and a shifting of intentional focus from the referents to awareness per se. As mentioned above, a referent can be a visual image and include sounds, light, fragrances, meaning, and the internal luminous energies that are connected with the internal perception of the embodied ego.

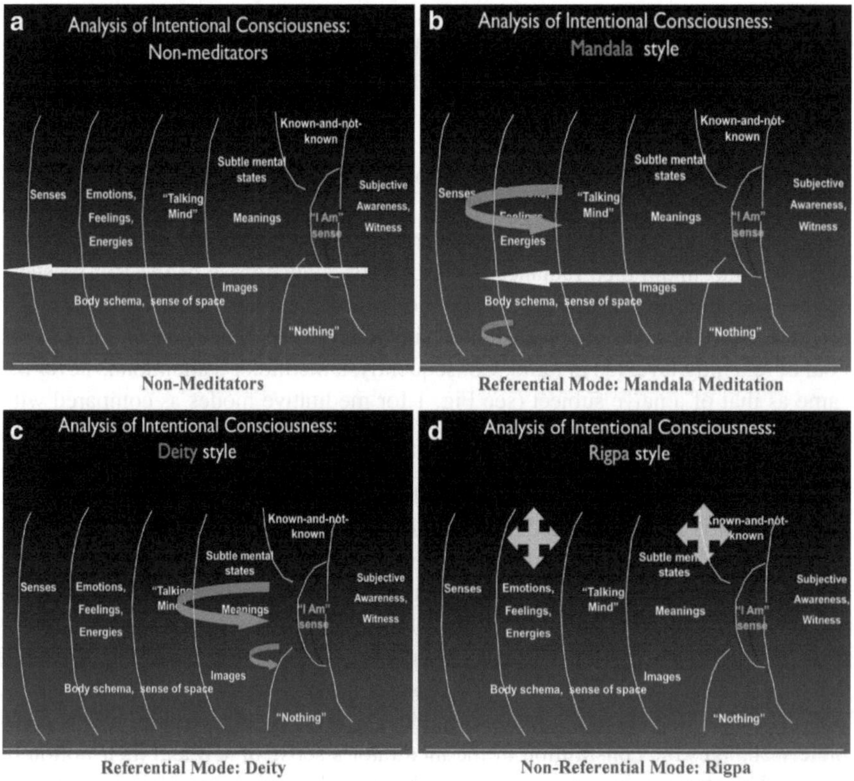

Fig. 1 Modes of intentional consciousness in different types of meditation: (**a**) Non-meditators, (**b**) Referential mode: Mandala meditation, (**c**) Referential mode: Deity Yoga meditation, (**d**) Non-referential mode: Rig-pa meditation. Phenomenological structures of intentionality in non-meditators (**a**), Tibetan Mandala meditation (**b**), Deity Yoga meditation (**c**), and *Rig-pa* meditation (**d**). Structures were obtained through interviewing subjects in the focus groups concerning their experience of the self. The layering of experience is a result of introspective attention directed at the self in an egological mode. In the ordinary state, consciousness is "about" its objects, i.e., intentionality is directed at the objects. In Mandala and Deity Yoga, consciousness is also resting on the objects; however, it is introspective (intra-subjective), as opposed to the complete externality to the self, or inter-subjectivity in the ordinary state. In Deity Yoga, the focus is closer to the empirical center of the self than in Mandala meditation. In *Rig-pa* meditation, consciousness is not focused or intended on anything, it is an open space of presence, which equally relates to all content including itself

Our group of participants included meditators who practiced visual meditations in referential and non-referential modes ($n>35$), meditations in non-referential mode (n approx. 20), and a mixed group of Vipashyana (n approx. 20). We also had informants who were not meditators ($n>300$). All participants were interviewed with regard to their experience of self-awareness in the practice. Two types of visual meditation, mandala and Deity Yoga, meet the criteria for the referential and non-referential modes. The Mandala practice focuses mostly on the mentally imagined environment of the deity; the Deity Yoga includes one's bodily identification with a deity as a part of visualization, which is more of a self-referencing and less an object-referencing mode. Figure 1 shows the phenomenological maps of consciousness in these different kinds of meditation. Vipashyana practice is not included as the group was not homogenous.

It is possible to distinguish three levels of transformation in Tibetan meditation: (1) object-oriented, allocentric consciousness, e.g., mandala meditation; (2) a self-centering process, e.g., Deity Yoga; and (3) transcendence of the internal self-consciousness into a gestalt of the transcendental ontological ground of "expansive openness and undimmed radiance," i.e., objectless awareness of the Rig-pa state. While many Buddhist practices work predominantly on level (1) or level (2), in any practice, from the start, the practitioner is instructed to be alert yet relaxed, with the goal of reaching level (3) in mind. Consequently, a meditator's attention is never the same as that of a naïve subject (see Fig. 1 for meditative modes as compared with non-meditative modes). A beginning meditator is introduced to the concept and empirical possibility of the pure awareness and is instructed to seek this state. In this manner, the meditator's awareness is differentiated from other mental processes from the start, and the subsequent visualization training takes place against the background of already modified cognitive processes. In the growing one-pointedness of the mind and the gradual reversal of attention toward its phenomenological origin in pure awareness, the meditator's phenomenological self-construct is completely dismantled. Existence of the self now consists of the empirical existence of the pure awareness with its ecstatic (transcending any state) perceptual intensity. Initially purely metaphysical, pure awareness is realized experientially. This completely changes the constitution of the meditator's sense of self and their notion of reality. The changes of consciousness in these modes are so persistent and empirically evident that they must have the cognitive correlates.

Translation of Phenomenological Findings into Cognitive Psychological Experiment. In order to design an experiment on visual cognition, researchers had to interpret the phenomenological findings in terms of cognitive psychology. In Table 2, we present the analysis of the phenomenologically distinct types of Tibetan meditation among our research practitioners. Vipashyana,[39] a phenomenologically mixed group, was included in analysis because as a transitional practice between the referential and non-referential modes, it served as a control group.

For the cognitive experiment, we selected the practices that rendered both the strongest phenomenological differences and the strongest distinctions in the cognitive tasks. These practices were Deity Yoga and Rig-pa meditations.

Designing the Experiment. Since it was unknown whether the assumed cognitive changes take place only during meditation or carry over into everyday consciousness, the researchers tested the influence of meditation on visual and spatial processing both at the baseline and after the meditation session (Table 3).

As explained above, Rig-pa or the non-referential mode of consciousness is both a specific practice and a goal of meditation. The referential mode of consciousness in Deity Yoga is not a goal in itself, but is a way of attainment of the non-referential

Table 2 Crossover between cognitive psychological variables and phenomenologically distinct types of meditation

Cognitive psychological category	Phenomenologically distinct types of meditation			
	Mandala	Deity Yoga	Vipashyana	Rig-pa
Cognition	+	+	+–	–
Attention	+++++	++++++	++	+–
Awareness	+	+	+	++++++++
Body schema	+	++	+–	–
Body image	–	+++	+–	–
Affect	+++	+++	-	+–
Sense of self	–	+++++++++	+–	–
Type of imagery	Allocentric	Egocentric	None or random	None
Performance on computerized imagery/visuospatial working memory tasks (pretest vs. posttest)	No effect	Distinct consistent increase of visual-spatial processing	Mixed responses	No effect

Note. Results of the analytic juxtaposition of different types of Tibetan meditation that were identified phenomenologically (see Fig. 1), with the spectrum of psychological parameters (modes of the natural attitude), which can be cognitively tested. Vipashyana is a mixed group representing a transitional state of meditation, from witnessing the contents of the mind (referential mode), to abiding in the witnessing consciousness per se without reference to its contents, i.e., *Rig-pa*. These kinds of meditation were also subjected to the pilot testing with computerized working visuospatial memory tasks, in order to gain a preliminary assessment of performance. Table 2 demonstrates that the experience of meditation involves cognitive modalities differently. Even though the presence or absence of modality in the subjective experience does not necessarily exactly correspond to the actuality of cognitive functioning, we cannot fully exclude this possibility. Therefore, we chose to compare the kinds of meditation that display the highest degree of difference in juxtaposition with subjectively perceived psychological parameters, i.e., *Rig-pa* and Deity Yoga

Table 3 The schema of cognitive-phenomenological research design

Stage in experiment			
Experimental group	Pretest	Impact	Posttest
Non-meditators, rest	Visual testing Spatial testing	Rest	Visual testing Spatial testing
Deity Yoga meditators	Same	Meditation Deity Yoga	Same
Rig-pa meditators	Same	Meditation Rig-pa	Same
Non-meditators, ordinary imagery control	Same	Maintaining unspecified image	Same

Note. Represents the experimental design that was derived from the data of cognitive-phenomenological mapping. Two selected meditation groups, Deity Yoga and Rig-pa, were accompanied by the control groups of non-meditators at rest, and non-meditators attempting to maintain an ordinary image. All groups were measured twice, before and after the impact, which consisted of the meditation session for the Rig-pa and Deity Yoga groups, at rest, and maintaining an unspecified image for non-meditators

mode, or Rig-pa. Therefore, in a single meditation session of Deity Yoga, the practice begins with visualization and after a period of time resolves into the pure awareness. These two states of experience may involve different brain networks.[40] Consequently, it was important to clearly isolate in the experiment the stage of the stable mental imagery in Deity Yoga, before the resolution of the image into Rig-pa begins. This was accomplished through the timing of the experimental meditation session.

The Deity Yoga meditation consists of generating a vivid, elaborate visualization of a chosen deity and her or his entourage and temporarily replacing one's image of oneself and the surrounding world with that generated image. It involves generation, maintenance, and transformation of the image. After the preliminary prayers and contemplations that provide the overall context for the meditation practice, the meditator volitionally suspends the habitual cognitive processes which support internal perception of self and the world and becomes empty of all cognitions. From this emptiness arises a point of light, and from that, a seed syllable, which transforms into a mandala of the chosen deity, including the practitioner in the form of the deity. The elaborateness of this visualization is believed to facilitate the development of the ability of open awareness. The meditator maintains visualization of himself or herself as a deity in the palace-like world, with numerous identical deities extending in all directions from the self and forming an infinite three-dimensional net of deities. This visualization is thought to help the meditator understand the mental constitution of the subject-object division and of the self-world relation. The meditator then visualizes rainbow-colored lights emanating from all of the deities, filling the entire space, and absorbing into their body (which is now that of the deity). The lights confer blessings and empower desirable qualities of being such as compassion, wisdom, goodness, and strength, as well as an overall state of enlightenment. Next, the meditator visualizes that they are absorbing all of the lights and outer deities into the self and then absorbing the deity's body into a seed syllable at the center of the chest space, then into a point of light, and finally into an empty space, until there is nothing left. In that emptiness of the mind, open nonconceptual awareness (Rig-pa) dawns. The last stage of Deity Yoga meditation usually involves dedicating the practice to the benefit of all sentient beings.

It takes a fairly good meditator about 40 minutes to enter the state of Rig-pa. In our experiment, meditators were interrupted for testing after 20 minutes, when the mental image is still intact, present, and before Rig-pa dawns.

Toward Future Research: Phenomenological Analysis Versus Meditation Styles

All groups performed the same on the baseline (pretest). After 20 minutes of meditation, Deity Yoga practitioners demonstrated a dramatic increase in performance on visual and spatial working memory tasks compared with the other groups. From the standpoint of the theory of visual cognition, these results suggest that Deity Yoga does not affect the day-to-day VWM, but trains a capacity to access heightened

Table 4 Phenomenological-cognitive mapping continued: Constituents of Deity Yoga and Rig-pa meditation

Constituent of the experience	Rig-pa	Deity Yoga (at the stage of generation of the deity)
Variety in the contents of the meditation	None or very little	Introspective exercise is modified according to the personal traits of the meditator, with the goal of a greater benefit to their particular mind
Presence of thoughts	Ideally, none; however, random thoughts can arise	Yes; controlled thoughts are specifically generated (the thoughts about the deity, their nature, etc.)
Presence of images	Ideally, none; however, random images may arise	A variety of complex, multidimensional visual images
Presence of existential assumptions (what is real, one's identity, life and death)	Possible, but limited	Yes, but thoughts are not owned; they belong to the imagined deity
Presence of emotions	None, except for the foundational fullness of the empty mind	Yes; affect is specifically cultivated, dependent on the character of the deity (e.g., joy or anger), and used to release latent (and sometimes socially prohibited) emotions
Verbal processes, semantic processing	Expected to be absent	May be present, dependent on the stage of Deity Yoga meditation session
States related to qualia (defined qualities of sensory perception)	None; ideally the state is supposed to be qualia-free	Yes; various qualities are specifically generated and controlled
Phenomenological attitude (i.e., the relationship between the perceiving/cognizing conscious subject and the field of the phenomena perceived)	Phenomenon-free	The subject is immersed in a phenomenal field and "moves" with it
Concentration/focusing	Absent, except for the initial stages of practice, when the focus can be on the awareness, which is the subject of meditation itself	Centrally used as one of the main faculties engaged in the practice
Direction of attention	Evenly hovering, or rather turned into pure awareness; in the initial stages of practice, attention can be reversed onto the source of attention itself	From the subject to the internal objects, then intrasubjectively, and accompanied by a reverse flow
Individual identity	Complete cancellation of individual identity	Identity is sustained, but altered

(continued)

Table 4 (continued)

Constituent of the experience	Rig-pa	Deity Yoga (at the stage of generation of the deity)
Depth and dimensionality of the internal space	Ideally, the space construct is transcended and absent	Generates multidimensional, dynamic spaces
Sense of time	Cancelled	Altered
Sense of the self and notion of the self	Absent	Present, though embedded in image(s) and modified, except in brief periods of transition from image to image
Locus of transformational changes	No transformational changes	The experientially perceived individual self
Imagination	Absent	Fully present and cultivated, including visualization and dynamic mental simulation
Awareness	Nonconceptual	Conceptual
Body image	Absent	Cultivated and modified
Contents of cognition	No cognition except for self-aware, non-intentional awareness	Intentionally generated and controlled cognitive experiences
Overall assessment	Experientially transcends mental functioning	Experientially engages and refines mental functioning

Note. Presents a comparison between the constituents of experience in *Rig-pa* and Deity Yoga meditation. By comparison with the ordinary state of experience this is modified in both kinds of meditation, but differently. This difference in experience corresponds to the difference in visuospatial processing, which was discovered experimentally (Kozhevnikov et al., 2009); however, differences in cognitive changes between these two types of meditation may include other parameters, not only those that were included in the experiment

visual-spatial resources, possibly by engaging the executive attention network that governs the allocation of brain resources within the working memory unit. If we continue phenomenological-cognitive mapping (Table 4), it expands the possible explanations of the discovered effect.

Table 4 clearly shows that the actual experience in meditation contains many phenomenological and cognitive distinctions that are not taken into the account in the simple categorization of the Deity and Rig-pa meditation as the two different meditation styles: concentration (referential mode) and open awareness (non-referential mode). While such differentiation is enough for the purposes of enlightenment in Tibetan Buddhism, it may be not enough to understand all the brain mechanisms involved in the generation of experience in both practices. The phenomenological picture in the analysis of the actual description is much richer than the classification into the two distinct meditation styles proposed.

First, Rig-pa is empty of cognitive contents (i.e., no images, thoughts, or emotions), but not of cognition itself. The contents of cognition are different: the content of the Deity Yoga generation stage involves bringing forth a complex, meaningful mental image of the body and the ornaments of a deity, the mandala (environment), the affects, and actions belonging to the deity. Second, while Rig-pa involves evenly

distributed attention, not directed toward anything in particular, attention in Deity Yoga visualization is focused and progresses from an object-referent to a subject-referent mode. Third, there is a significant difference in the perception of time and space. While the sense of time is absent in Rig-pa, it is present in Deity Yoga meditation; the concept of space is absent, or exists in a seed form in Rig-pa, while a Deity Yoga meditator generates a multidimensional internal space. Finally, the sense of self, including the body, is absent in Rig-pa and present in Deity Yoga. All these components of experience have their own brain correlates which undergo the modification in two types of meditation and can contribute to the enhanced visual processing.

Besides uncovering the fact that visual processing can be trained, our experiment established that the mental imagery indeed utilizes the networks of the brain associated with normal visual processing, as would be in the case of computer games. However, if meditation includes more than the visual modalities of internal imagery, e.g., semantic processing or body schema, the involvement of executive attention will include subsystems responsible for these different modalities and be more nuanced than in the case of pure mental visual imagery.[41] The process will also be different from an earlier described "switching off" of the irrelevant networks for the maintenance of focused internalized attention and inhibition of inappropriate information (Cf. Aftanas & Golocheikine, 2002). If Deity Yoga were exclusively to train the inhibition of irrelevant networks and the reallocation of resources,[42] other cognitive functions might have been depleted, but phenomenologically, they are not. Meditators, having rich multidimensional internal experience which they can self-regulate, make one assume that they find extra sources to replenish the energy of their neural networks. In such a complex, yet repetitively trained process, optimization of the allocation of resources may not be limited to the functioning of the top-down attentional regulation, but may involve the dynamic systemic restructuring of the whole global space of the brain.[43] With the executive attention network being causal in the top-down regulation of visual memory (Dehaene & Naccache, 2001; Zanto, Rubens, Thangavel, & Gazzaley, 2011), these two options may be connected.[44] The engagement of the global workspace, supporting phenomenological open awareness, may be a necessary prerequisite for the training of the attentional network, as is customary in the Buddhist rDzogs-chen practice mode. This nonconceptual mental state appears to have its own neural correlates, as in the phenomenologically analogous Zen meditation, in which semantic processing is inhibited (Pagnoni, Cekic, & Guo, 2008).

First-person self-reporting, with the consequent second-person phenomenological analysis of the data, proposes that separate experiments need to be conducted to clarify the formation of the nonconceptual state. In her previous phenomenological studies of the nonconceptual state in Vedantic and Hesychastic types of meditation, the present researcher showed that the nonconceptual state is not just a state of mental awareness but that it emerges in connection with the dynamics of the bodily awareness and engages the components of phenomenological somatic constitution.[45] Therefore, if there is a representation of this state in the brain, it should involve the somatosensory complex, right hemispheric areas associated with embodied emotions, and the insulae.[46]

With regard to the shifts from the referential to non-referential mode, this shift may also involve a switch in neuro-assemblies supporting these two different modes of experience. Phenomenologically, meditation which actualizes the non-referential mode appears to be similar to, but not the same as, childhood experiences of undifferentiated awareness. Thus, the potential change in functional connectivity can be analogous to the developmental segregation of local regions and integration of distant regions into subnetworks (Vogel, Power, Petersen, & Schlaggar, 2010).

While enormous progress has been made to identify the neural correlates of consciousness, crucial aspects of it are still very contradictory. Some researchers suggest that both focused attention and awareness are supported by the same localized attentional networks, while others suggest that attention and awareness are not a unitary phenomenon, but can be fluid with different brain locations involved.[47] These locations range from local networks to the global working space of the brain (Van Heuvel & Hulshoff, 2010). Phenomenological isolation of attention and awareness, through the juxtaposition of different modes of meditation in experimental designs, can contribute to the understanding of this aspect of the mind. Visual Deity Yoga meditation offers a possibility for such experiential isolation of the conditions of visual focused attention and open awareness. The above phenomenological juxtaposition, as well as differences discovered in cognitive processing between mature Rig-pa and early stages of Deity Yoga meditation, calls for more research which can shed light on whether attention and awareness are mediated by the same or different networks of the brain.

Phenomenological analysis and phenomenological-cognitive mapping appear necessary for future research. Cognitive phenomenology, indeed, creates a set of constraints for both the theory and understanding of experience, raising questions with regard to the application of cognitive theories in cases of experience which are not taken in the natural attitude, such as meditation. This research also raises questions concerning meaningful existential and religious connotations of the imagery. Meditators generate images based on oral and textual descriptions, not on pictures – what does this tell us about the participation of the brain in mental imagery? These questions require the phenomenological analysis of experience to be linked to scientific experimentation.

Notes

1. For more on the adequacy of introspective self-reporting, see Marti, Sackur, Sigman, and Dehaene (2010).
2. For more on the cognitive science's attitude towards experience, see Barinaga (2003).
3. For more on engaging the researcher's experience in the cognitive experiment, see Varela, Thompson, and Rosch (1991). For scientific views on the relationship between mental states and consciousness, see Leisman and Koch (2009) and Zeman (2001). For an example of the empiricist, commonsense analysis of experience, see Kosslyn (1995) and Kosslyn et al. (1990). For an example of correlations between the experience of meditation and the metabolism of the brain, see Baerentsen et al. (2010).

4. For more on problems in first-person reporting of mental states, see Lutz, Lachaux, Martinerie, and Varela (2002). For more on verbalization of the ineffable, see Sells (1994).
5. For more on constraints in neurophenomenological designs, see Petitot, Varela, Pachoud, and Roy (1999).
6. For an example of how the analysis of experience eliminates excessive theorizing, see Kinsbourne (2006).
7. For differences in the resting state in meditators and non-meditators, see Tei et al. (2009).
8. For more on changes of perception in spiritual development, see Louchakova (2007a).
9. For more on mental imagery, see Thomas (2010). For an example of early studies of meditation as a relaxation technique, see Pfeiffer (1966).
10. For more on visual working memory, see Cowan (2001), Luck and Vogel (1997), Mitchel and Cusack (2011), and Zhang and Luck (2008). For separate processing of visual and spatial aspects of imagery, see Kosslyn (1994) and Kravitz, Saleem, Baker, and Mishkin (2011). For cognitive testing, see Metzler and Shepard (1974). For training of the spatial component of visual memory, see Erfani and Erfanian (2004).
11. Compare Carruthers and Veillet (2012), who argue that conscious thoughts do not co-constitute pre-reflective experience. Thus, reporting may not be able to reach the pre-reflective level of consciousness, which would make it impossible to study meditation phenomenologically. However, research of religious or spiritual experience shows that articulation reaches as deeply as the bodily components of meditation. For the phenomenology of meditative experience, see Ales Bello (2009), Louchakova (2005), and Louchakova and Warner (2003).
12. For more on the influences of meditation on the contents of consciousness, see Louchakova (2005) and Louchakova-Schwartz (2011). For more on pure subjectivity, see Forman (1990), Fasching (2008), Wallace (1999, 2000), and Zahavi (2002).
13. For more on the importance of contexts of meditation, see Lindberg (2005), Mayo (2009), Newberg (2006), and Wachholtz and Pargament (2005).
14. For the spontaneous activity of the brain in the resting state, see Doucet et al. (2011) and Jang et al. (2011). For more on the intrinsic state, see Biswal et al. (2010) and Mennes et al. (2011). For characteristic patterns of neural activity that support subjective experience in the resting state, see Lou et al. (1999) and Van den Heuvel and Hulshoff Pol (2010).
15. For changes of experience in meditators, see Louchakova (2005). For an example of changes of brain metabolism during a single meditation session, see Newberg, Pourdehnad, Alavi, and d'Aquili (2003) and Wang et al. (2011). For an example of attentional change in long-term meditation, see Chan and Woolacott (2007). For changes in neuroactivity in long-term meditators, see Huang and Lo (2009).
16. For meditation as a relaxation technique, see Pfeiffer (1966).
17. For connections between physiological and cognitive effects of meditation, see Schwartz, Davidson, and Goleman (1978) and Wenk-Sormaz (2005).
18. For meditation as an attentional strategy, see Fell, Axmacher, and Haupt (2010), Jha, Krompinger, and Baime (2007), Kozhevnikov et al. (2009), Lutz, Slagter, Dunne, and Davidson (2008), and Slagter, Davidson, and Lutz (2011). For more on top-down regulation, see Newberg (2010) and Newberg et al. (2003). For meditation as an attentional strategy, see Fell et al. (2010), Jha et al. (2007), Kozhevnikov et al. (2009), Lutz et al. (2008), and Slagter et al. (2011). For more on top-down regulation, see Newberg (2010) and Newberg et al. (2001, 2003).
19. For a review of findings about meditation, see Cahn and Polish (2006), Fell et al. (2010), and Neumann and Frasch (2006).
20. For an example of peaceful coexistence of conflicting theories in neuroscience, see Bechtel, Stufflebeam, Mundale, and Mandik (2001). For connections of specific cognitive function to the activities of the brain, see Benchenane, Tiesinga, and Battaglia (2011). For the localizationist and reticularist paradigms in neuroscience, see Bechtel and Stuffebeam (2001). For more on attention networks, see Fan et al. (2007). For the global space of the brain and awareness, see Raffone and Pantani (2010).
21. For theoretical models in neuroscience, see Bechtel et al. (2001).

22. For examples of correlations between the state of meditation experience and neuroactivity, see Aftanas and Golocheikine (2001), Banquet and Lesèvre (1980), Fell et al. (2010), Jang et al. (2011), Lo, Huang, and Chang (2003), Lutz, Greischar, Rawlings, Richard, and Davidson (2004), Rubik (2011), and Travis and Shear (2010).
23. For more on the connections between experience and neuroplasticity, see Rosenzweig (2003). For neuroplasticity connected with meditation, see Brefczynski-Lewis, Lutz, Schaefer, Levinson, and Davidson (2007), Chiesa (2010), Fell et al. (2010), Green and Turner (2010), Lazar et al. (2005), and Xiong and Doraiswamy (2009).
24. For approaches to understanding experience, see Camic, Rhodes, and Yardley (2003) and Frie (2003). For more on the natural attitude, see Luft (2011).
25. For the distinction between the natural attitude and phenomenological analysis, see Finlay (2008). For an example of rigorous studies of meditation experience, see Pekala, Wenger, and Levine (1985) and Travis, Arenander, and DuBois (2004).
26. For examples of the internal states, see Lutz et al. (2008).
27. On phenomenological ontology, see Dillon (1997).
28. For more on the differences between the experience of consciousness in the natural attitude and phenomenological awareness of the structures of consciousness, see Louchakova-Schwartz (2011).
29. For more on the effects of the different objects of concentration, see Goleman (1988).
30. This analysis of Buddhist meditation is based on the author's observations, collected over more than 20 years of meditation training with Sogyal Rinpoche, at the Nyingma Institute in Berkeley, in the Vajrayana Center in Russia, as well as teaching meditation as a mandated teacher of various forms of Tantra, Advaita Vedanta, Hesychasm, and Sufism. For more, see Louchakova (2005) and Kungurtsev and Louchakova (1994).
31. For more on the state of pure subjectivity in Vedantic meditation, see Louchakova-Schwartz (2012). For more on the advanced stages of meditation, see Bader (1990) and Guenther (1992).
32. For more on Husserl's approach to the study of consciousness, see Toadvine (2002). For more on transcendental subjectivity, see Zahavi (2005). For more on pure consciousness, see Forman (1990).
33. For more on non-reductive physicalism, see Murphy (1999) and Wildman and Brothers (1999).
34. For the spectrum, see Travis et al. (2004).
35. For an example of texts used for this analysis, see Guenther (1992), Longchenpa (1975), Stcherbatskoi (1903), and Wangual (1993).
36. For an example of reflective analysis used by Merleau-Ponty, see Merleau-Ponty (1962).
37. Compare Husserl's notion of intentionality in Ideas I (1980).
38. For more on reconstitution of the self in meditation or spiritual practice, see Louchakova (2005, 2007a).
39. Vipashyana (mindfulness, Sanskrit term. Analogous practice in Theravada Buddhism is rendered as Pali Vipassana) consists of witnessing the various elements of the mind. The advanced level of practice, in which attentional focus on the elements of self-construction resolves into a state of insight, is believed to be one of the five aspects of Rig-pa per se. With regard to the advancement of the non-referential mode, Vipashyana is a more advanced practice than Deity Yoga at the stage at which we interrupted it for the experiment, but less advanced than the full Rig-pa.

    Rig-pa is the condition of the open, dynamic "ecstatic intensity" of pure being/nonconceptual awareness, devoid of objects. The Rig-pa state has aspects and degrees, and the practice of Rig-pa has developmental stages. We were not able to account in our work for the fine distinctions between developmental stages within the category of Rig-pa practitioners, since accomplished Rig-pa meditators are very hard to find. However, since the results of the tests for the Rig-pa group were closely clustered, we believe that the discrimination between the practices in our work was sufficient to match the sensitivity of our visual-spatial tasks. For more details on this, as well on the tests used, see Kozhevnikov et al. (2009). For the first report of the layering of experience in egological consciousness, see Louchakova (2005).

40. For the neuro-correlates of referential and non-referential modes of consciousness with regard to the self, see Josipovic (2010), and Travis et al. (2004).
41. For an example of the nuanced involvement of the prefrontal cortex/executive attention network in different types of meditation, see Davanger, Ellingsen, Holen, and Higdahl (2010), and Fujiki et al. (2008).
42. For the different manner in which resources can be allocated between modules of working memory, see Morey, Cowan, Morey, and Rouder (2011).
43. For an example of non-localizationist theories of visual consciousness, see Baars (1999).
44. For more on the connection between executive attention and awareness, see Dehaene and Naccache (2001).
45. For somatic component in the state of pure awareness, see Louchakova (2005, 2007a, 2007b, 2007c, 2008).
46. For the implicit subjectivity, see Schore (2010). For more on the areas of the brain involved with the constitution of embodied self-experience, see Chen et al. (2010), and Craig (2010).
47. For a localizationist account of attention/consciousness, see Chan and Woolacott (2007), Posner and Dehaene (1994), Posner and Petersen (1990), and Taylor and Fragopanagos (2007). For dynamic networks and global workspace accounts, see Freeman (2009), Leisman and Koch (2009), Lundervold (2010), and Raffone and Srinivasan (2009).

References

Aftanas, L. I., & Golocheikine, S. A. (2001). Human anterior and frontal midline theta and lower alpha reflect emotionally positive state and internalized attention: High resolution EEG investigation of meditation. *Neuroscience Letters, 310*(1), 57–60.

Aftanas, L. I., & Golocheikine, S. A. (2002). Non-linear dynamic complexity of the human EEG during meditation. *Neuroscience Letters, 330*, 143–146.

Ales Bello, A. (2009). The divine in Husserl and other explorations. *Analecta Husserliana: Vol. 98*. New York, NY: Springer.

Baars, B. J. (1999) Attention vs. consciousness in the visual brain: Differences in conception, behavior, neuroanatomy, and physiology. *Journal of General Psychology, 126* (3), 224–233.

Bader, J. (1990). Meditation in Shankara's Vedanta. New Delhi, India: Aditya Prakashan.

Baerentsen, K. B., Stødkilde-Jørgensen, H., Sommerlund, B., Hartmann, T., Damsgaard-Madsen, J.,... Green, A. C. (2010). An investigation of brain processes supporting meditation. *Cognitive Processing, 11*(1), 57–84.

Banquet, J. P., & Lesèvre, N. (1980). Event-related potentials in altered states of consciousness. *Progress in Brain Research, 54*, 447–453.

Barinaga, M. (2003). Buddhism and neuroscience. Studying a well trained mind. *Science, 302*, 5642–5644.

Bechtel, W., & Stuffebeam, R. S. (2001). Epistemic issues in procuring evidence about the brain. In W. Bechtel, R. Stufflebeam, J. Mundale, & P. Mandik (Eds.), *Philosophy and the neurosciences: A reader* (pp. 55–81). Malden, MA: Blackwell.

Benchenane, K., Tiesinga, P. H., & Battaglia, F. P. (2011). Oscillations in the prefrontal cortex: A gateway to memory and attention. *Current Opinion in Neurobiology*. Advance online publication. doi:10.1016/j.conb.2011.01.004

Biswal, B. B., Mennes, M., Zuo, X. N., Gohel, S., Kelly, C., Smith, S. M.,...Milham, M. P. (2010). Toward discovery science of human brain function. *Proceedings of the National Academy of Sciences of the United States of America, 107*(10), 4734–4739.

Brefczynski-Lewis, J. A., Lutz, A., Schaefer, H. S., Levinson, D. B., & Davidson, R. (2007). Neural correlates of attentional expertise in long-term meditation practitioners. *Proceedings of the National Academy of Sciences, 104*(27), 11483–11488.

Cahn, B. R., Delorme, A., & Polish, J. (2010). Occipital gamma activation during Vipassana meditation. *Cognitive Process, 11*, 39–56.
Cahn, B. R., & Polich, J. (2006). Meditation states and traits: EEG, ERP, and neuroimaging studies. *Psychological Bulletin, 132*(2), 180–211.
Camic, P. J., Rhodes, J., & Yardley, L. (Eds.), *Qualitative research in psychology: Expanding perspectives in methodology and design* (pp. 243–273). Washington, D.C.: American Psychological Association.
Carruthers, P., & Veillet, B. (2012). The case against cognitive phenomenology. In T. Bayne & M. Montague (Eds.), *Cognitive phenomenology* (pp. 35–55). New York, NY: Oxford University Press.
Chan, D., & Woolacott, M. (2007). Effects of level of meditation experience on attentional focus: Is the efficiency of executive or orientation networks improved? *The Journal of Alternative and Complementary Medicine, 13*(6), 651–657.
Chen, T. L., Babiloni, C., Ferretti, A., Perrucci, M. G., Romani, G. L., Rissini, P. M.,...Del Gratta, C. (2010). Effects of somatosensory stimulation and attention on human somatosensory cortex: an fMRI study. *Neuroimage, 53*(1), 181–188.
Chiesa, A. (2010). Vipassana meditation: Systematic review of current evidence. Journal of *Alternative and Complementary Medicine, 16*(1), 37–46.
Cowan, N. (2001). The magical number 4 in short-term memory: A reconsideration of mental storage capacity. *Behavioral Brain Science, 24*, 87–118.
Craig, A. D. (2010). The sentient self. *Brain Structure and Function, 214*(5–6), 563–77.
Davanger, S., Ellingsen, O. Holen, A., & Higdahl, K. (2010). Meditation-specific prefrontal cortical activation during acem meditation: An fMRI study. *Perception & Motor Skills, 111*(1), 291–306.
Davidson, R. J., Goleman, D. J., & Schwartz, G. E. (1976). Attentional and affective concomitants of meditation: A cross-sectional study. *Journal of Abnormal Psychology, 85*(2), 235–238.
Dehaene, S., & Naccache, L. (2001). Towards a cognitive neuroscience of consciousness: Basic evidence and a workspace framework. *Cognition, 79*(1–2), 1–37.
Dillon, M. C. 1997. *Merleau-Ponty's ontology*. 2nd ed., Northwestern University studies in phenomenology and existential philosophy. Evanston, Ill: Northwestern University Press.
Doucet, G., Naveau, M., Petit, L., Delcroix, N., Zago, L., Crivello, F.,...Joliot, M. (2011). Brain activity at rest: A multi-scale hierarchical functional organization. *Journal of Neurophysiology*. Advance online publication. doi:10.1152/jn.00895.2010
Erfani, A., & Erfanian, A. (2004). The effects of mental practice and concentration skills on EEG brain dynamics during motor imagery using independent component analysis. *Conference Proceedings of the IEEE Engineering in Medicine and Biology Society, 1*, 239–42.
Fan, J., Byrne, J., Worden, M. S., Guise, K. G., McCandliss, B. D., Fossella, J., & Posner, M. I. (2007). The relation of brain oscillations to attentional networks. *The Journal of Neuroscience, 27*(23), 6197–6206.
Fasching, W. (2008). Consciousness, self-consciousness, and meditation. *Phenomenology and the Cognitive Sciences, 7*, 463–483.
Fell, J., Axmacher, N., & Haupt, S. (2010). From alpha to gamma: Electrophysiological correlates of meditation-related states of consciousness. *Medical Hypothesis, 75*(2), 218–224.
Finlay, L. (2008). A dance between the reduction and reflexivity: Explicating the "Phenomenological Psychological Attitude." *Journal of Phenomenological Psychology, 39*, 1–32.
Forman, R. K. C. (1990). Introduction: Mysticism, constructivism and forgetting. In R. K. C. Forman (Ed.), *The problem of pure consciousness* (pp. 3–49). New York, NY: Oxford University Press.
Freeman, W. J. (2009). Vortices in brain activity: Their mechanism and significance for perception. *Neural Networks, 22*, 491–501.
Frie, R. (Ed.). (2003). *Understanding experience: Psychotherapy and postmodernism*. New York: NY: Routledge.

Fujiki, M., Akiyoshi, J., Yoshida, T., Tabata, M., Kabsawa, H., Kobayashi, H., & Shimomura, T. (2008). Functional brain mapping during recitation of Buddhist scriptures and repetition of the Namu Amida Butsu; A study in experienced Japanese monks. *Turkish Neurosurgery, 18*(2), 134–141.

Gallagher, S. (1997). Mutual enlightenment: Recent phenomenology in cognitive science. *Journal of Consciousness Studies, 4*(3), 195–214.

Goleman, D. (1988). *The meditative mind: The varieties of meditative experience*. New York, NY: Penguin, Putman.

Green, R., & Turner, G. (2010). Growing evidence for the influence of meditation on brain and behavior. *Neuropsychological Rehabilitation, 20*(2), 306–311.

Guenther, H. (1992). *Meditation differently*. Delhi, India: Motilal Banarsidas.

Huang, H. Y., & Lo, P. C. (2009). EEG dynamics of experienced Zen meditation practitioners probed by complexity index and spectral measure. *Journal of Medical Engineering and Technology, 33*(4), 314–321.

Husserl, E. (1980). Ideas pertaining to a pure phenomenology and to a phenomenological philosophy. *Collected works Edmund Husserl*. The Hague, Boston, Hingham, MA, USA: M. Nijhoff.

Jang, J. H., Jung, W. H., Kang, D. H., Byun, M. S., Kwon, S. J., Choi, C. H., & Kwon, J. S. (2011). Increased default mode network connectivity associated with meditation. *Neuroscience Letters, 487*(3), 358–62.

Jha, A. P., Krompinger, J., & Baime, M. J. (2007) Mindfulness training modifies subsystems of attention. *Cognitive, Affective and Behavioral Neuroscience, 7*(2), 109–119.

Josipovic, Z. (2010). Duality and nonduality in meditation research. *Consciousness and Cognition, 19*(4), 1119–1121.

Kinsbourne, M. (2006). From unilateral neglect to the brain basis of consciousness. *Cortex, 42*, 869–874.

Kosslyn, S. M. (1995). Mental imagery. In S. M. Kosslyn & D. N. Osherson (Eds.), *Visual cognition: An invitation to cognitive science* (Vol. 2) (pp. 267–296). Cambridge, MA: The MIT Press.

Kosslyn, S. M., Seger, C., Pani, J. R., & Hillger, L. A. (1990). When is imagery used in everyday life? A diary study. *Journal of Mental Imagery, 14*(3&4), 131–152.

Kozhevnikov, M., Louchakova, O., Josipovic, Z., & Motes, M. A. (2009). The enhancement of visuospatial processing efficiency through Buddhist deity meditation. *Psychological Science, 20*(5), 645–653.

Kravitz, D. J., Saleem, K. S., Baker, C. I., & Mishkin, M. (2011). A new neural framework for visuospatial processing. *Nature Reviews Neuroscience, 4*, 217–230.

Kungurtsev, I., & Louchakova, O. (1994). The unknown Russian mysticism: Pagan sorcery, Christian Yoga, and other esoteric practices in the former Soviet Union. *Gnosis, 32*, 20–27.

Lazar, S. W., Kerr, C. E., Wasserman, R. H., Gray, J. R., Greve, D. N., Treadway, M. T.,...Fischl, B. (2005). Meditation experience is associated with increased cortical thickness. *Neuroreport, 16(*17), 1893–1897.

Leisman, G., & Koch, P. (2009). Networks of conscious experience: Computational neuroscience in understanding life, death, and consciousness. *Reviews in the Neurosciences, 20*, 151–176.

Lindberg, D. A. (2005). Integrative review of research related to meditation, spirituality, and the elderly. *Geriatric Nursing, 26*(6), 372–377.

Lo, P. C., Huang, M. L., & Chang, K. M. (2003). EEG alpha blocking correlated with perception of inner light during Zen meditation. *The American Journal of Chinese Medicine, 31*(4), 629–642.

Longchenpa. (1975). *Kindly bent to ease us. Part one: Mind* (H. V. Guenther, Trans.) Berkeley, CA: Dharma Publishing.

Lou, H. C., Kjaer, T. W., Friberg, L., Wildschiodtz, G., Holm, S., & Nowak, M. (1999). A 15O-H2O PET study of meditation and the resting state of normal consciousness. *Human Brain Mapping, 7*, 98–105.

Louchakova, O., & Warner, A. (2003). Via Kundalini: Psychosomatic excursions in transpersonal psychology. *The Humanistic Psychologist, 31*(2–3), 115–158.

Louchakova, O. (2005). Ontopoiesis and union in the prayer of the heart: Contributions to psychotherapy and learning. In A. T. Tymieniecka (Ed.), *Analecta Husserliana: The yearbook of phenomenological research (Vol. 91). Logos of phenomenology and phenomenology of the logos. Book four: The logos of scientific interrogation. Participating in nature –life- sharing in life* (pp. 289–311). Dordrecht, UK: Kluwer.

Louchakova, O. (2007a). Ontopoiesis and spiritual emergence: Bridging Tymieniecka's phenomenology of life and transpersonal psychology. In A.-T. Tymieniecka (Ed.), *Analecta Husserliana: Phenomenology of life- from the animal soul to the human mind* (Vol. 94) (pp. 43–68). Dordrecht, UK: Kluwer.

Louchakova, O. (2007b). Prayer of the Heart, ego-transcendence and adult development. *Existential Psychoanalysis, 18*(2), 261–287.

Louchakova, O. (2007c). Spiritual heart and direct knowing in the Prayer of the Heart. *Existential Psychoanalysis, 18*(1), 81–102.

Louchakova, O. (2008). Reconstitution of the self during the study of Advaita Vedanta (Traditional Indian teachings of non-dual consciousness): Phenomenological investigations. In *Toward a science of consciousness*, April 8–12, 2008, Tucson AZ, Consciousness Research Abstracts (315), p. 194.

Louchakova-Schwartz, O. (2011). Self and world: Vedanta, Sufism and Presocratics in phenomenological view. In A.-T. Tymieniecka (Ed.), *Phenomenology/ontopoiesis retrieving cosmic horizons of antiquity* (pp. 423–438). Dordrecht, Germany: Springer.

Louchakova-Schwartz, O. (2012). Intuition of life in Tymieniecka's Phenomenology with a reference to intuition of Sat in Śaṅkara's Advaita Vedanta. *Culture & Philosophy: A Journal for Phenomenological Inquiry, 2011*, 40–60.

Luck, S. J., & Vogel, E. K. (1997). The capacity of visual working memory for features and conjunctions. *Nature, 390*, 279–281.

Luft, Sebastian. 2011. *Subjectivity and lifeworld in transcendental phenomenology*. Evanston, Ill: Northwestern University Press.

Lundervold, A. (2010). On consciousness, resting state fMRI, and neurodynamics. *Nonlinear Biomedical Physics, 4* (Suppl. 1), S1–S9.

Lutz, A., Brefczynski-Lewis, J., Johnstone, T., & Davidson, R. J. (2008). Regulation of the neural circuitry of emotion by compassion meditation: Effects of meditative expertise. *Public Library of Science One, 3*(3), e1897.

Lutz, A., Dunne, J. D., & Davidson, R. J. (2007). Meditation and the neuroscience of consciousness: An introduction. In P. Zelazo, M. Moscovitch, & E. Thompson (Eds.), *The Cambridge handbook of consciousness* (pp. 499–554). New York, NY: Cambridge University Press.

Lutz, A., Greischar, L. L., Rawlings, N. B., Richard, M., Davidson, R. J. (2004). Long-term meditators self-induce high-amplitude gamma synchrony during mental practice. *Proceedings of the National Academy of Sciences, 101*(46), 16369–16373.

Lutz, A., Lachaux, J. P., Martinerie, J., & Varela, F. J. (2002). Guiding the study of brain dynamics by using first-person data: Synchrony patterns correlate with ongoing conscious states during a simple visual task. *Proceedings of the National Academy of Science, 99*(3), 1586–1891.

Lutz, A., Slagter, H. A., Dunne, J. D., & Davidson, R. J. (2008). Attention regulation and monitoring in meditation. *Trends in Cognitive Science, 12*(4), 163–169.

Lutz, A., & Thompson, E. (2003). Neurophenomenology: Integrating subjective experience and brain dynamics in the neuroscience of consciousness. *Journal of Consciousness Studies, 10* (9–10), 31–52.

Marbach, E. (2006). No heterophenomenology without autophenomenology: Variations on a theme of mine. *Phenomenology and the Cognitive Sciences, 6*(1–2), 75–87.

Marti S., Sackur, J., Sigman, M., & Dehaene, S. (2010). Mapping introspection's blind spot: reconstruction of dual-task phenomenology using quantified introspection. *Cognition, 115*(2), 303–313.

Mayo, K. R. (2009). Support from neurobiology for spiritual techniques for anxiety: A brief review. *Journal of Health Care Chaplaincy, 16*(1–2), 53–7.

Mennes, M., Zuo, X. N., Kelly, C., Di Martino, A., Zang, Y. F., Biswal, B.,...Milham, M. P. (2011). Linking inter-individual differences in neural activation and behavior to intrinsic brain dynamics. *Neuroimage, 54*(4), 2950–2959.

Merleau-Ponty, M. (1962). *Phenomenology of perception* (C. Smith, Trans.). London, UK: Routledge and Kegan Paul. (Original work published 1945)

Metzler, J., & Shepard, R. N. (1974). Transformational studies of the internal representation of three-dimensional objects. In R. Solso (Ed.), *Theories in cognitive psychology: The Loyola Symposium* (pp. 147–201). Hillsdale, NJ: Erlbaum.

Mitchel, D. J., & Cusack, R. (2011). The temporal evolution of electromagnetic markers sensitive to the capacity limits of visual short-term memory. *Frontiers in Human Neuroscience, 5*, 18.

Morey, C. C., Cowan, N., Morey, R. D., & Rouder, J. N. (2011). Flexible attention allocation to visual and auditory working memory tasks: Manipulating reward induces a trade-off. *Attention, Perception, & Psychophysics, 3*(2), 458–472.

Murphy, N. (1999). Introduction. In R. Russel, N. Murphy, T. Meyering, & M. Arbib (Eds.), *Neuroscience and the person. Scientific perspectives on divine action* (pp. i-xxxv). Berkeley, CA: Center for Theology and Natural Science.

Nagel, T. (1971). Brain bisection and the unity of consciousness. *Synthese, 22*, 396–413.

Neumann, N.-U., & Frasch, K. (2006). The neurobiological dimension of meditation–Results from neuroimaging studies. *Psychotherapie Psychosomatik Medizinische Psychologie, 56*, 488–492.

Neumann, N.-U., & Karel, F. (2006). The neurobiological dimension of meditation - results from neuroimaging studies. *Psychotherapie Psychosomatik Medizinische Psychologie, 5*(12), 488–492.

Newberg, A. B. (2006). Religious and spiritual practices: A neurochemical perspective. In P. NcNamara (Ed.), *Where God and science meet: How brain and evolutionary studies alter our understanding of religion* (pp. 15–32). Westport, CT: Praeger.

Newberg, A. B. (2010). Neurobiology of meditation. In D. Monty & B. Beitman (Eds.), *Integrative psychiatry* (pp. 339–358). Oxford, NY: Oxford University Press.

Newberg, A., Alavi, A., Baime, M., Pourdehnad, M., Santanna, J., & d'Aquili, E. (2001). The measurement of regional cerebral blood flow during the complex cognitive task of meditation: a preliminary SPECT study. *Psychiatry Research, 106*, 113–122.

Newberg, A., Pourdehnad, M., Alavi, A., & d'Aquili, E. G. (2003). Cerebral blood flow during meditative prayer: Preliminary findings and methodological issues. *Perceptual and Motor Skills, 97*, 625–630.

Overgaard, M. (2004). On the naturalising of phenomenology. *Phenomenology and the Cognitive Sciences, 3*, 365–379.

Pagnoni, G., Cekic, M., & Guo, Y. (2008). Thinking about not-thinking: Neural correlates of conceptual processing during Zen meditation. *Public Library of Science One, 3*(9), e3083.

Pekala, R., Wenger, C., & Levine, R. (1985). Individual differences in phenomenological experience: States of consciousness as a function of absorption. *Journal of Personality and Social Psychology, 48*(1), 125–32.

Petitot, J., Varela, F. J., Pachoud, B., & Roy J-M. (Eds.). (1999). *Naturalizing phenomenology: Issues in contemporary phenomenology and cognitive science*. Stanford, CA: Stanford University Press.

Pfeiffer, W. M. (1966). Concentrative self-relaxation by exercises derived from Buddhistic respiration-meditation and from respiration therapy. Zeitschrift für Psychotherapie und *Medizinische Psychologie, 16*(5), 172–81.

Posner, M. I., & Dehaene, S. (1994). Attentional networks. *Trends in Neuroscience, 17*, 75–79.

Posner, M. I., & Petersen, S. E. (1990). The attentional system of the human brain. *Annual Review of Neuroscience, 13*, 25–42.

Raffone, A., & Pantani, M. (2010). A global workspace model for phenomenal and access consciousness. *Consciousness and Cognition, 19*(2), 580–596.

Raffone, A., & Srinivasan, N. (2009). An adaptive workspace hypothesis about the neural correlates of consciousness: Insights from neuroscience and meditation studies. *Progressive Brain Research, 176*, 161–180.

Raffone, A., & Srinivasan, N. (2010). The exploration of meditation in the neuroscience of attention and consciousness. *Cognitive Processing, 11*(1), 1–7.

Rosenzweig, M. R. (2003). Effects of differential experience on the brain and behavior. *Developmental Neuropsychology, 24*(2&3), 523–540.

Rubik, B. (2011). Neurofeedback-enhanced gamma brainwaves from the prefrontal cortical region of meditators and non-meditators and associated subjective experiences. *Journal of Alternative and Complementary Medicine, 17*(2), 109–115.

Schore, A. (2010). The right brain implicit self: A central mechanism of the psychotherapy change process. In J. Petrucelli (Ed.), *Knowing, not-knowing and sort of knowing: Psychoanalysis and the experience of uncertainty* (pp. 177–202). London, UK: Karnac.

Schwartz, G. E., Davidson, R. J., & Goleman, D. J. (1978). Patterning of cognitive and somatic processes in the self-regulation of anxiety: Effects of meditation versus exercise. *Psychosomatic Medicine, 40*(4), 321–8.

Sells, M. A. (1994). *Mystical languages of the unsaying*. Chicago, IL: University of Chicago Press.

Slagter, H. A., Davidson, R. J., & Lutz, A. (2011). Mental training as a tool in the neuroscientific study of brain and cognitive plasticity. *Frontiers in Human Neuroscience, 10*, 5–17.

Stcherbatskoi (Stcherbatsky, Sherbastky), T. (1903). *Teoriya poznaniya i logika po ucheniyu pozdneishikh buddistov* [*Theory of knowledge and Logic According to Later Buddhists*]. St. Petersburg, Russia: Izd. Fak. Vost Yazykov Imp. Sankt-Peterburgskogo Universiteta. No. 14.

Taylor, J. G., & Fragopanagos, N. (2007). Resolving some confusions over attention and consciousness. *Neural Networks, 20*(9), 993–1003.

Tei, S., Faber, T. L., Lehmann, D., Tsujiuchi, T., Kumano, H., Pascual-Marqui, R. D.,...Kochi, K. (2009). Meditators and non-meditators: EEG source imaging during resting. *Brain Topography, 22*(3), 158–165.

Thomas, N. J. T. (2010). Mental imagery. In E. N. Zalta (Ed.), *The Stanford encyclopedia of philosophy*. Retrieved from http://plato.stanford.edu/archives/fall2010/entries/mental-imagery/

Toadvine, T. (2002). Leaving Husserl's cave? The philosopher's shadow revisited. In T. Toadvine, & L. Embree (Eds.), *Merleau-Ponty's readings of Husserl* (pp. 71–94). Netherlands: Kluwer.

Travis, F., Arenander, A., & DuBois, D. (2004). Psychological and physiological characteristics of a proposed object-referral/self-referral continuum of self-awareness. *Consciousness and Cognition, 13*, 401–420.

Travis, T., & Shear, J. (2010). Focused attention, open monitoring and automatic self-transcending: Categories to organize meditations from Vedic, Buddhist and Chinese traditions. *Consciousness and Cognition, 19*(4), 1110–1118.

Travis, F., & Wallace, R. K. (1999). Autonomic and EEG patterns during eyes-closed rest and transcendental meditation (TM) practice: The basis for a neural model of TM practice. *Conscious and Cognition, 8*(3), 302–318.

Van den Heuvel, M. P., & Hulshoff Pol, H. E. (2010). Exploring the brain network: A review on resting-state fMRI functional connectivity. *European Neuropsychopharmacology, 20*(8), 519–534.

Varela, F. J., Thompson, E., & Rosch, E. (1991). *The embodied mind: Cognitive science and human experience*. Cambridge, MA: The MIT Press.

Vogel, A. C., Power, J. D., Petersen, S. E., & Schlaggar, B. L. (2010). Development of the brain's functional network architecture. *Neuropsychology Review, 20*(4), 362–375.

Wachholtz, A. B., & Pargament, K. I. (2005). Is spirituality a critical ingredient of meditation? Comparing the effects of spiritual meditation, secular meditation, and relaxation on spiritual, psychological, cardiac, and pain outcomes. *Journal of Behavioral Medicine, 28*(4), 369–384.

Wallace, B. A. (1999). The Buddhist tradition of Samatha. Methods of refining and examining consciousness. In F. Varela, & J. Shear (Eds.), *The view from within. First-person approaches to the study of consciousness* (pp. 175–187). Thorverton, UK: Imprint Academic.

Wang, D. J., Rao, H., Korczykowski, M., Wintering, N., Pluta, J., Khalsa, D. S., & Newberg, A. B. (2011). Cerebral blood flow changes associated with different meditation practices and perceived depth of meditation. *Psychiatry Research, 191*(1), 60–67.

Wangual, T. (1993). *Wonders of the natural mind. The essence of Dozgchen in the native Bon tradition of Tibet*. Barryton, NY: Station Hill.

Wenk-Sormaz, H. (2005). Meditation can reduce habitual responding. *Alternative Therapies in Health and Medicine, 11*(2), 42–58.

Wildman, W. J., & Brothers, L. A. (1999). A neurophysical-semiotic model of religious experience. In R. Russel, N. Murphy, T. Meyering, & M. Arbib (Eds.), *Neuroscience and the person. Scientific perspectives on divine action* (pp. 347–417). Berkeley, CA: Center for Theology and Natural Science.

Wittig, A. F. (2001). *Introduction to psychology*. New York, NY: McGraw-Hill.

Xiong, G. L., & Doraiswamy, P. M. (2009). Does meditation enhance cognition and brain plasticity? *Annals of the New York Academy of Sciences, 1172*, 63–69.

Zahavi, D. (2002). Anonymity and first-personal givenness: An attempt at reconciliation. In D. Carr & C. Lotz (Eds.), *Subjektivität – Verantwortung – Wahrheit* (pp. 75–89). Frankfurt am Main, Germany: Peter Lang.

Zahavi, D. (2005). *Subjectivity and selfhood*. Cambridge, MA: The MIT Press.

Zahavi, D. (2010). Naturalized phenomenology. In S. Gallagher & D. Schmicking (Eds.), *Handbook of phenomenology and cognitive science* (pp. 3–19). Dordrecht, UK, New York, NY: Springer.

Zanto, T. P., Rubens, M. T., Thangavel, A., & Gazzaley, A. (2011). Causal role of the prefrontal cortex in top-down modulation of visual processing and working memory. *Nature Neuroscience, 14*, 656–661.

Zeman, A. (2001). Consciousness. *Brain, 124*(7), 1263–89.

Déjà Vu: William James on "The Brain and the Mind," 1878 – A Comment on Current Trends in Neurophenomenology Defining the Application of James's Radical Empiricism to Psychology

Eugene Taylor

It has seemed to me that the 6 hours that the trustees of the Lowell Fund have done me the honor to place at my disposal, could not be better spent than in taking a single subject, the brain, and seeing exactly how much recent investigations have explained its action and particular how much they may be said to have cleared up or made less mysterious the action of consciousness in each one of us.

(William James, Manuscript Lectures, 1988, p. 31)

Let us, for purposes of discussion, define neurophenomenology as the attempt to reconcile first-person science with third-person science and the enactive approach as the idea that cognition does not stand alone but is inextricably interconnected with the physical body and its interaction with the environment.[1] The dialogue concerning the relation of first-person science to third-person science began with the articulation of autopoiesis in the work of Maturana and Varela (1987), the idea in the new biology that self-governing systems that have emergent properties, and progressed to Varela's enunciation of a means by which objectivist, third-person accounts of phenomena in traditional cognitive science might be linked to first-person accounts of immediate experience, when personal experience is systematically examined as it has been in the traditional discipline of phenomenology, as

E. Taylor, Ph.D. (✉)
School of Psychology and Interdisciplinary Inquiry, Saybrook University, Harvard Medical School, 747 Front Street, 3rd Floor, San Francisco, CA 94111, USA

Department of Psychiatry and Massachusetts General Hospital, Harvard Medical School, Boston, MA, USA
e-mail: etaylor@igc.org; susan.gordon@snet.net

S. Gordon (Ed.), *Neurophenomenology and Its Applications to Psychology*, DOI 10.1007/978-1-4614-7239-1_4,

defined by those who follow the works of Edmund Husserl and Maurice Merleau-Ponty (Varela, Thompson, & Rosch, 1991; Varela & Shear, 1999). Essentially, we are defining here the central problem now confronting the neuroscience revolution, namely, the relation between the brain and the mind or put another way the relation between our objective understanding of a thing and our subjective phenomenal experience of it in the immediate moment—what James referred to as "knowledge about" versus direct "acquaintance with" (James, 1890, v. 2, p. 221). Dubbed today the so-called Hard Problem, the inherent nature of the neuroscience revolution speaks to interdisciplinary communication between fields that have not been in conversation before, such as cognitive neuroscience, brain neurophysiology, evolutionary psychology, and the philosophy of mind, among others. On the other hand, in a Jamesean sense, it also poses the probably unsolvable problem of how science can objectively study itself. In my opinion, it cannot, the entirely subjective decision to do science in the first place being a sentiment, according to James.

Neuroscientists in this regard like to think that they are working at the cutting edge of a revolution focused on the biology of consciousness and have yet to confront the phenomenology of the science making process itself. It would certainly come as a surprise for many to find that these arguments had already been taken up by philosophers and psychologists more than 130 years ago and actually dealt with at certain points, in my opinion, in a much more sophisticated way. We may take William James as a relevant example, and the construct of radical empiricism embedded in his tripartite metaphysics. Central to understanding radical empiricism for the point we would like to make here is its application to contemporary psychology. The difficulty here is that most psychologists take psychology as a branch of the social sciences, while the type of psychology James was getting at is a phenomenologically oriented, intuitively driven psychology in the immediate moment.

As a case study we may derive parts of this understanding from James's various Lowell Lectures. Over the course of his career, James gave three series, each one important in its own right. The first series was given in 1878, as "The Brain and the Mind," drafts of which were later incorporated into *The Principles of Psychology* (1890). The second series, given in 1896, was on "Exceptional Mental States." These represented a transition from a cognitively oriented psychology of consciousness in *The Principles* (1890) to a dynamic psychology of subliminal states of consciousness by the mid-1890s (Taylor, 1982/2010a).[2] These lectures then became foundational to major chapters in his *Varieties of Religious Experience* (1902). Then there was a final series given in 1906 as "Pragmatism" (1907), which he intended as one of the major texts of his tripartite metaphysics of pragmatism, pluralism, and radical empiricism, an attempt to sum up his final metaphysical position on personality, consciousness, and immediate experience.

For the present, we will confine ourselves to the Lowell Lectures of 1878, which are instructive for certain contents in *The Principles of Psychology* of interest to contemporary neurophenomenologists. Within them, among other early sources, we also find the germ of James's later epistemology of radical empiricism.

The 1878 Lowell Lectures on "The Brain and the Mind"

The Lowell Institute had been launched in 1836 by the Lowell family, a scion of Boston society that had built its fortunes through development of textile mills in Lowell, Massachusetts. The Lowell Institute, which had its own lecture hall down on Boylston Street in Boston's Back Bay, sponsored public lectures delivered by experts in their respective fields. They offered the equivalent of a year's salary for a professor to give a single short series of six or eight public lectures. The only provision was that the field of expertise of the speaker itself had to be addressed and the most current knowledge being described had to be made understandable to the common men and women of Boston's working class (Weeks, 1966).

James had begun his career in 1873 in physiology, teaching to undergraduates Prof Jeffries Wyman's old course in anatomy and physiology in Harvard College. This came under the aegis of the Department of Natural History. He was much more interested in mental science, however, having already staked a claim in the 1860s on the role of consciousness within the scheme of Darwin's theory of natural selection (Taylor, 1990). Thus, after he began teaching, he had petitioned Pres. Eliot for a position in philosophy, should one open up. Meanwhile, the one and only time he considered a teaching career somewhere other than Harvard and, likely as a ploy to prod Eliot into changing his appointment, James accepted an invitation from Pres. Daniel Coit Gilman at Johns Hopkins to give a short course of lectures, which James understood to be a job application. He wrote to Gilman that the subject would be "in a way expository and critical of recent speculations regarding the connection between the mind and the body" (James, 1988, p. 448).

At Hopkins, he gave ten lectures with an emphasis on the brain and the senses. The University reports that the title of the series was "The Senses and the Brain and their Relation to Thought."[3] James repeated them for the Lowell Institute a few months later but had to modify the series somewhat when Augustus Lowell, the executor of the Lowell Institute, proposed he shorten the series to only six. The combined content of both versions then found a major place in his article "Are We Automata?" (1879a) and later, as we have said, in chapters of his *Principles of Psychology* (1890). At the same time, James and his medical colleagues, James Jackson Putnam and Henry Pickering Bowditch, had been conducting the first investigations in brain neurophysiology in the USA in Bowditch's laboratory at the Harvard Medical School, and out of these endeavors, James had assembled the first laboratory in experimental psychology in America for student instruction in 1875.[4] It was located in the Lawrence Scientific School in Cambridge and, also in 1875, began in conjunction with a new course taught by James in physiological psychology, the first course in physiological psychology to be taught in an American university.

Meanwhile, James and Charles Sanders Peirce, Chauncey Wright, Nicholas St. John Green, and others began to meet, holding monthly meetings of the so-called Metaphysical Club, which alternated between the dining rooms of the Peirce and James households, out of which later came the American philosophical movement called pragmatism. Along with the lectures Peirce gave that became the germ of the

Graduate School, the philosophy of science generated out of these meetings contributed to the development of the experimentally oriented mental sciences at Harvard from the 1870s onward in physiology, neuropathology, and experimental psychology. Attacking the mind/body problem head on, these were also the influences that caused James in *The Nation* in 1876 to argue that physiologists should study philosophy and philosophers should study physiology to approach the functional relation of the mind to the body as a whole (James, 1876). This was the tack James took in his Lowell Lectures of 1878.

Writing to Lowell:

> The course I propose is on the subject to which I have given the most attention—The senses and the Brain and their relation to Mind. I feel confidant of my ability to make the course both entertaining to the many by its abundance of new facts and illustrations and instructive to the few by its theoretic conclusions. I think I can safely say that I am neither a materialistic partisan nor a spiritualistic bigot. (James, 1988, p. 451)

On The Brain and the Mind

The great problem we face, James announced, in opening the lectures to his Boston audience, is the spectacular advance over the previous decade in our knowledge of the brain, suggesting a new understanding of the mind in the field of metaphysics but at the same time raising suspicions of a brutal materialism lurking like an idol behind the brilliant veil when it came to the actual relation between the mind and the body. What we hear more and more in science, he said, is all about the body and nothing about the mind. James indicated he was going to try to bridge this gap in six public lectures, in which he would present what new knowledge we had about the brain and try to integrate this with what occurs in consciousness when we attempt to look within and introspect upon our experience (James, 1877).

His lectures were laid out in such a way that at the core he presented material on the physiology of the brain, making liberal use of lantern slides comparing the brains of lower animals to that of a human brain. He interspersed this comparison with commentary on the philosophical implications of this work, which in the end amounted to an argument against materialism and an affirmation of love, hope, feelings, and aspirations as just as important to the whole picture of the person and not to be left out.

Lecture One

James began with the claim that:

> In the new Darwinian age, scientists, particularly physiologists, assert how recent advances have cleared up all our questions, nay banished the mystery which used to hang about the action of the mind, and constituted a new psychology which exploded and rendered obsolete the old views of mental action, all based on a priori speculation and metaphysics. (James, 1988, p. 437)

Meanwhile, the detractors of the new science, as he had said earlier, saw the most brutal materialism lurking behind what its adherents called enlightenment and scientific progress, like some hideous heathen idol whose form is dimly seen through the glare of fireworks and golden dust and dazzling vapors of incense with which its followers continually fill the air before it. James's aim, however, was to unite the yaysayers and the naysayers.

He then drew a brain and a spinal cord on the blackboard, identified the afferent and efferent nerves, and then compared this to the brain and cord of the lamprey, dog, turtle, bird, beaver, horse, porpoise, elephant, and monkey.[5] In this, he was able to demonstrate where the human brain appeared to be so much more complex with the addition of the cerebral hemispheres. He then said that what we know physiologically, however, is a mere objective account of what we can see, which does not explain the internal experience of these systems at work. Their subjective form is only guessed by analogy or by generating a fictitious anatomy (James, 1874).

Lecture Two

James then turned to different kinds of cells, showing slides of cellular structure in the brain and cord. Reminding his audience that still we do not know how the cells are connected, he told them.[6] We have groups of cells and bundles of nerve fibers, but physiology is invented by the experts at this point trying to figure out the mechanism of how excitements travel from one cell to another. Lots of theories, but nothing definite. Perhaps, he conjectured, the excitement may go right across the separate cells. In making up the explanations, we naturally turn to our own subjective experience to fill in what we do not know. We know the subjective facts and the anatomical facts, but not the link between them, except that up to now, all we hear from the scientists is the denigration of all things subjective, when the physiological evidence remains slim and the better explanation has already come to us through introspection, which the new pundits claim is an obsolete method. Each discipline erects its barriers and stays focused only on its own subject matter. We as proprietors of a body and a mind must feel the insufficiency of leaving out the one or the other.

Lecture Three

James opened by reminding his auditors that, having established that animals become more complex as we go up the evolutionary scale, their brains become more complex, particularly in terms of the need for larger and larger cerebral hemispheres. The bottom layers seem to govern the more primitive aspects of motion, balance, sensation, reflexes, the instincts, and so on, built up in layers until the instincts give way to habits, providing economy for the hemispheres to operate without having to remember to breathe or perform other automatic tasks geared toward survival.

Behaviors that have been learned can then operate automatically. The hemispheres themselves negotiate language, cognition, a sense of time for past, present, and future, nay, self-consciousness, and are naturally more evolved in man than in lower animals. The hemispheres allow us to act from ideas instead of stimuli, and ideas presuppose a storing up of experiences.

Lecture Four

James opened the fourth lecture by summarizing the third one, mentioning memory as a common property, the hemispheres providing so many more tracts, whereas in other animals the tendency is one track for all acquisitions. He then launched into a description of experiments by Flourens, Hitzig and Ferrier, and Monk.[7] In the motor regions, destruction causes pigeons to wink and dodge and dogs to show defensive movements of the head. As well there is a paralysis of volitional movements. As far as the auditory and optic centers were concerned, ablation is accompanied by a mental blindness, where the organism sees, but does not understand. With regard to sensations, what is not felt is supplied by the mind. His conclusion was that each motor and sensory element has its home. When that is destroyed, representation is wiped out. The remaining representations are then not enough to constitute thought.

When he turned to humans, James discussed aphasia, gestures, and language disturbance and referred to evidence for the localization of functions in the brain. This led him to a comparison with phrenology, in which modern theories of the brain at least corroborate the pigeon hole theory put forward in that system of thinking.[8]

Lecture Five

The succession of our thoughts might be explained physiologically as the nerve currents running through the body, throwing offshoots here and there except that there may be alternate explanations possible. James then entered again onto a more in-depth comparison between the physiological theory of localization of function and the theory behind phrenology. Phrenology presumes that within each bump on the head sits a little homunculus guiding the destiny of the person from that trait. Philoprogenitiveness suggests a person who loves children in one fell swoop, with all the complicated assumptions that that must entail. Physiologists, on the other hand, try to understand the physical systems, from elementary to complex, building atomistically from the bottom up, through cells, structures, and systems, and thence thought and belief that might lead to the conclusion that a person loved children. Such an explanation does not detract from the usefulness of phrenology as an art of character development, at the same time that physiologists cannot automatically presume that physical processes explain the phenomenon and that is all that is necessary.

He then launched into a discussion of the inherent philosophy of materialism accompanying the scientific view, which he maintained was actually unnecessary for the explanation to remain scientific. In addition, the youthful nature of the field precluded rules that were too fixed as there were numerous examples where functions were not so localized or occurred temporarily and then not at that place again. The materialist viewpoint pretends it is a philosophy, however, and in that disguise pervades physiology. It presumes that anything mental is a by-product of something physiological—that the higher is always reducible to the lower and the lower is the explanation for it. However, there is nothing to preclude a physiology based on the opposite—that mental events cause physiological changes to take place and that conscious activity and all that that entails is the cause of the underlying physiology. Materialism is not a rational explanation for everything. "Thus the charge of degrading the high by deducing it from the low which these authors escape in the theoretical sphere is a charge which they cannot escape in the physical sphere" (James, 1988, p. 23).

Lecture Six

James opened his last lecture by restating the contemporary theory in vogue that thoughts follow each other in certain order because brain tracks awaken each other in that order. Once we know everything about the nervous system, the pseudo-theory of the scientific materialists goes; we will be able to explain all actions in human history. Feelings, in this context, are considered inert, uninfluential, a mere passenger. Their philosophy is such that "since frog's legs twitch and automatic habitual acts may be performed by men without consciousness and yet outwardly appear rational, then when rational acts appear with consciousness, this consciousness is unconcerned with their production" (1988, p. 24). This, James said, is nothing more than an unproven hypothesis. The fact is, the way science proceeds is that the true difference is in aesthetics, what the perceiver desires to be true as far as the finest and most elegant explanation is concerned. Then, there is a search for the facts to prove such sentiments.[9]

But:

> If consciousness can load the dice, can exert a constant pressure in the right direction, can feel that nerve processes are leading to the goal, can reinforce & strengthen these & and the same time inhibit those which threaten to lead astray, why, consciousness will be of invaluable service.... But the nature of consciousness as we know it is just this and nothing else. In the first place, it feels the goal, it knows the interests, it is the sole standard of use. Apart from some consciousness staking, posting, creating some particular end, as good [meaning a sufficient explanation,] we cannot talk about the brain being useful or efficient, its actions being appropriate, at all [meaning it is not the brain acting but the person]. (James, 1988, p. 26)

He went on to summarize Darwin, whose model of consciousness considers survival as the *summum bonum* and then measures the utility of every creature's organs by these ends. Such a reaction of the nervous system is efficient, such a habit, fit, and so on. As such, Darwin estimates whichever act is worthy of approbation or encouragement. Interpreting James somewhat here, we see that there may be no God driving this selection, but Darwin certainly introjects his own conclusions into the process. And, if Darwin can do it, why not other animals?

James's point seemed to be that if we survey the field of consciousness from lowest to highest, we find that "it always seems to be comparing and selecting." He said, "Consciousness in short, at each of its stages is busy picking out from the data offered by its lower stage, some one item, noticing, emphasizing, attending to, pursuing it and ignoring all the rest" (1988, p. 27).

In this there is a hierarchy. Sensation is the lowest, perception the next stage; reasoning is the next higher faculty, aesthetic experience the next:

> In ethics, choice reigns supreme. The choice in the moral life is between the interests: Which is my true interest, to be rich or to be just? To be powerful or to be pure? To escape pain or at any cost of pain to do service to humanity. In fact, we choose one, out of several equally possible future selves. (1988, p. 27)

Consciousness, in short, is an educator of the nervous system. And to say that it is not, James declared, is absurd.

Finally, James ended the lectures on that same note:

> Many persons nowadays seem to think that any conclusion must be very scientific if the arguments in favor of it are all derived from twitching frog's legs—especially if the frogs are decapitated—& that on the other hand, any doctrine chiefly vouched for by the feelings of human beings—with heads on their shoulders, must be benighted and superstitious.... They seem to think, too, that any vagary or whim, however unverified, of a scientific man must needs form an integral part of science itself; that when Huxley for example, has ruled feeling out of the game of life, and called it a mere bystander, supernumerary, the matter is settled.... I know nothing more deplorable than this undiscriminating gulping down of everything materialistic as peculiarly science. (1988, p. 29)

His concluding statement was:

> I, for one, as a scientific man & a practical man alike, deny utterly that Science compels me to believe that my consciousness is an *ignis fatuus* or outcast, & I trust that you, too, after the evidence of this evening will go away strengthened in the natural faith that your delights & sorrows, your loves and hates, your aspirations & efforts are real combatants in life's arena, & not impotent, paralytic spectators of the game. (1988, p. 29)

A Comment on the Relevance of the Lectures of 1878

The arguments for materialism as the only legitimate science are as relevant today as in James's time. According to James, however, the claims to absolute truth are largely mere guesses based on the expression of rationality as a sentiment. Instead, consciousness shows at every level that choices are being made that lead to the survival of the organism, not just genetically, not just over the lifetime or the history of an organism, but also right now in the immediate moment, where physiology goes on as well as rational striving and aesthetic choice. But only a primitive physiology was permitted entry into the domain of scientific investigation in James's time, when, according to James, the very facts of our experience of consciousness contradict such a claim. It is then important to understand how James brought these ideas forward with regard to a scientific understanding of consciousness.

The Evolution of James's Definition of Consciousness

The 1878 Lowell Lectures were delivered during a period in James's career when he had already established his place within the circle of Darwinists at Harvard as the man who was most interested in the role of consciousness, and particularly conscious choices, in the process of natural selection. Among his earliest publications, in a review in the *Nation*, James maintained that individual conscious choices impact the social evolution of the species, which in turn alter the biological evolution of human beings (James, 1865, 1868). He applied the principles of natural selection to the consciousness of geniuses, as a case in point. Most of us live habitually in a state of unreflective thoughtlessness. When we have to make choices, we decide on the basis of whatever it is that is at hand as far as our ideas are concerned. Life then goes on as it normally would. But the mind of the genius is in a constant state of ferment, full of possibilities that most of us have never thought of, and in this broiling sea, when a choice has to be made, it is always on the basis of a greater number of analogies that the genius has at hand. The products are entirely different than the norm, however. Whereas in the normal man or woman the choices are usually limited and narrow, selection nevertheless takes place. The genius, on the other hand, thinks in the midst of bounteous analogies and has a much greater selection to choose from. The result is possibilities that humans would not normally believe possible. When the genius penetrated through rumination an act of creation, it is often a product unlike any other. Yet once created, others see the possibility and now gain access into that domain of accomplishment. The genius breaks through, and the rest of us follow, continually alternating the landscape of the forces governing social evolution, which in turn have their effect on altering the course of biological survival, shaping it, improving upon it, and widening the possibilities of physical adaptation (James, 1988).

Consciousness as What Is at the Cognitive Center of Attention

While he argued for a more holistic view of the role of consciousness in the evolutionary process, James was willing to commit himself to positivistic reductionism as the basis for his two-volume *Principles of Psychology*. Other alternative approaches were presented by Associationism and also by Spiritualism, yet these were weak and unsystematic. There was, as yet, no other equally powerful epistemology to replace the positivist point of view. As far as his position in *The Principles* was concerned, psychology was the study of the correlation of brain states to mental states, and the thought was the thinker, nothing more. Therefore the focus of what constituted a psychology of consciousness was the object in the center of the field of attention.

Meanwhile, what was at the center of attention was an object that appeared to have continuity from moment to moment, yet it was embedded in a stream of consciousness that flowed ever onward. James borrowed the phrase, "the specious present" to describe the fact that "while nothing ever happens to us except in the immediate moment, yet, as the moment flows continually onward, it always bears some mark of both the past and the future with it" (James, 1890, v. 1, p. 609).

The Germ of His Later Radical Empiricism

James contracted to write the *Principles of Psychology* in 1878, believing it would take 2 years, when it actually took him 12. He declared as it finally went off to press that he was glad to get "that dropsical, tumescent mass" off his desk. James had struggled over which epistemology to follow in writing *The Principles* and finally decided to return to reductionistic positivism, the ground on which he had started. Many things had delayed him, not the least of which was his introduction to the French Experimental Psychology of the Subconscious and the launching of the British and American Societies for Psychical Research. Their scientific researches gave evidence that waking rational consciousness was not the only state there was but that state was where experimental psychology had confined itself. Altered states of consciousness were lurking beyond the margin of everyday awareness which, if proven to be true, mooted positivist epistemology as narrow and incomplete and hardly the standard by which to judge the entire spectrum of human experience. Nevertheless, he stuck with positivism in *The Principles*. He did mention, however, the work of the French, descriptions of hysteria, multiple personality, automatisms, and the like. While he tried to remain focused on reductionistic and positivistic science, nevertheless, these alternative epistemologies kept appearing.

We are given one provocative clue about these alternative epistemologies brewing in James's thinking when he cut down the two-volume, 1,200 page, *Principles* to the one volume, cut-and-paste version called *Psychology: Briefer Course* (1892), the textbook version that remained in use in American colleges for the next 20 years. In that shorter work, discussing the characteristics of the stream of consciousness, he left out one that had appeared prominently in *The Principles*, namely, that there was a world that existed outside, independent of the person.

The germ fully sprouted 2 years later when, during his address in 1894 as president of the American Psychological Association, he reminded his audience of the struggle he had been having with reductionistic positivism (James, 1895).

He had finally solved the problem and decided to throw it out altogether in favor of a new way of thinking about the assumptions underlying the way experimental science should be conducted, which in the preface to his *Will to Believe* in 1897, 5 years later, he finally gave the name radical empiricism. Experimental psychology was based solely on the rational ordering of sense data alone as a way to define all of reality, which he said, produced only an incomplete truth because of the limits of its epistemology. Empiricism had always been confined to data from the senses, which he believed was part of the problem. He proposed instead that empiricism stand for the full spectrum of human experience, not just sense perception alone. Understanding that this would shock his scientific audiences, he hoped to prepare them for this new explanation by affirming that he was an empiricist, but a radical one, in order to differentiate from the normal definition of the term.

Exceptional Mental States

By adopting such an assumption as the basis upon which experimental psychology could proceed, James launched into a flurry of new investigations throughout the 1890s in psychical research, experimental psychopathology, and the psychology of religion. One product of this thinking was his 1896 Lowell Lectures on Exceptional Mental States (Taylor, 2010a). His lecture titles were "Dreams and Hypnotism," "Automatism," "Hysteria," "Multiple Personality," "Demonical Possession," "Witchcraft," "Degeneration," and "Genius." The first four lectures established James' understanding of a dynamic psychology of subconscious states, applied largely to cases of psychopathology, while the second four demonstrated the workings of that same dynamic psychology within the individual, but now within the social sphere. Here, we see a major shift in his definition of consciousness. In *The Principles*, the thought was the thinker, and what psychology studied was the object at the center of the field of attention. That was the positivists' view. By 1896, however, as the germ of his radical empiricism began to grow, he redefined consciousness as "a field with a focus and a margin," where the margin controlled meaning. Every thought is accompanied by an emotion, he had declared in *The Principles*, which warms the thought and makes it our own. But emotion was so unstable, most of it being submerged in the subconscious. Nevertheless, if we talk about the stream of thought, we must always talk about the accompanying stream of feelings. They were the doorway into the mental imagery of the hypnogogic state, which in turn was the portal through which we experience the spectrum of states ranging from the psychopathological to the transcendent.

How to get at what was submerged in the subconscious, however? James's answer was hypnosis, automatic writing, and any other technique that would induce trance and lead to a dissociated state of consciousness. In these states we tap into a different strata of personal identity, one that can be identified by waking consciousness only to a certain extent, because they are made up of caches of interrelated memories within a given state but remain unknown to other states, especially waking consciousness. Following the theory of subliminal consciousness put forward by F. W. H. Myers, both dissolutive and evolutive states operate beyond the margin within each person. It was James's opinion that certain cases of multiple personality could occur where a superior personality might emerge and become the dominant person for the rest of the individual's life. It was then only a short step for investigators to understand that the therapist could appeal to this higher dimension of personality in therapy, the main goal of which could be not just recovery from the psychoneuroses, but the evolution of a higher and more integrated personality.

Within less than a year, James would announce Peirce's pragmatism to the philosophical world (James, 1898). It was the first formal airing of his own tripartite metaphysics of pragmatism, pluralism, and radical empiricism, but he made sure that he presented pragmatism as Peirce's idea. For James, it meant that beliefs were always tested by their consequences, and it was also a measure of how one can go about comparing different truth claims about the nature of ultimate reality to see if different seemingly contradictory explanations might not lead to the same ends. It caught fire nationally, along with Dewey and others, defining the Progressive Era in the United States for the first two decades of the Twentieth Century, before it went on to become the first uniquely American philosophy to have international consequences. Peirce disavowed James's interpretation, however, and claimed that all he meant was that to be perfectly logical one had to consider the consequences of the thought in question. See Taylor (2012a).

Consciousness as Transcendent Experience

An event of importance occurred in 1897 when James received the invitation to give the Gifford Lectures on Natural Religion at the University of Edinburgh. He accepted, but a year later endured an accident that severely weakened his heart. It was so bad he had to resort to injections of animal extracts directly into the chest. He cancelled the Gifford Lectures, but kept writing about pragmatism. Feeling better at one point, he reinstated them a year later, but did not finally deliver then until June of 1901 and 1902. Importantly, the content of the unpublished Exceptional Mental States Lectures of 1896 found a central place in his statement in *The Varieties of Religious Experience* (1902). Approaching his audience, he told them, "as a psychologist of religion," he declared that yes, his subject matter would be religion, by which he meant to confine himself to the interior experiences reported by the single individual. The ecclesiasticisms, the history of the priesthood, and the Holy Scriptures that had been written would not be his primary concern. Moreover, he declared his data base to be "the *documents humaine*." This was to be a phenomenological study using living human documents of personal experience.

Within an individual's interior experience, there were innumerable states of consciousness that were just then being uncovered by modern investigators, such as the psychical researchers in England, as well as neurologists and psychopathologists around Pierre Janet, Morton Prince, Boris Sidis, and Breuer and Freud. In short, exploration of subliminal or subconscious states, James believed, could lead to the wakening of mystical states of consciousness.

Mystical states of consciousness were transformative. They were ultimate in the sense of none higher. They could radically alter personality in the same way that a severe traumatic shock from an external source could. But their essential nature was that they represented the highest and most profound experiences human beings could report about their earthly existence. They were evolutive, as Myers had called them, reminiscent of what psychiatrist Richard Maurice Bucke believed—that human beings were moving toward such states as a species.

James gave four principal characteristics: they were noetic, in the sense that they imparted visionary knowledge; they were ineffable, in the sense that they were beyond words to describe; they were transient, in that they usually lasted but a few seconds or moments, yet their effects were felt from then on over a lifetime; and finally, often, they demanded complete surrender. There was no resistance possible. It was as if one was grasped and held there, as if giving oneself up to a higher power.

Though his work is seen by religious scholars such as Ann Taves (1999) as principally written within the Protestant tradition, James gave examples from a variety of world religions. Christianity, Islam, Hinduism, and Buddhism were all acknowledged, including anomalous sources such as New Thought and Theosophy. The nature of these states could come from any source, reputable or not. The important criteria were in terms of their "fruits for life." They were verified and confirmed according to whether or not they increased the moral and aesthetic character of daily living.

We may note three important points for our discussion here. First, James's analysis brought him to the conclusion that psychology's true contribution to the religious sphere lay along the lines of a cross-cultural, comparative psychology of mystical states of consciousness, studied from a phenomenological viewpoint. Second, among other definitions of radical empiricism, his primary focus was on pure experience in the immediate moment, before the differentiation of subject and object. While we can find numerous instances across the spectrum of experience that suggest such a state, certain kinds of mystical awakenings, at any rate, such as the apophatic type, might qualify as the most important examples.

At any rate, with the composition of *The Varieties* (1902), James had covered the entire spectrum of consciousness during his long career as a philosopher and psychologist. In 1865, he had established that consciousness was an active player influencing the course of natural selection, itself subject to those same laws. In 1890, following the positivistic trend of the times, he declared that consciousness was the scientific study of whatever cognition was at the center of the field of waking awareness. He modified those positions somewhat in 1892 when he eliminated the notion that the world of objects existed somehow independently of the person, and he then, making a clean break with reductionistic and materialistic positivism by 1894. This he articulated in his 1896 Lowell Lectures using a new model that consciousness was a field with a focus and a margin, where the margin, which controls the meaning of the cognitive thought, is in a constant state of expansion and contraction, sometimes now revealing more of the spectrum of states within the interior life of the person, sometimes restricting that view, making it look like as if only waking rational consciousness existed, or confining awareness pathologically on a single idea. Beyond the margin was the hypnogogic state, where abstract ideas were transformed into interior mental images, thereby giving access to a spectrum of interior states of consciousness ranging from the psychopathic to the transcendent. By *The Varieties* in 1902, he had come to the conclusion that the highest mystical states of consciousness that were possible for human beings to experience were the most profound with regard to the ultimate growth-oriented transformation of personality, aiming as they were toward the highest good human beings were capable of achieving.

After that, conceptually where was he to go? He had covered the full spectrum of experience with regard to the phenomena of consciousness. From that vantage point, he was forced to step back and ask if a science of consciousness was really possible. His tripartite metaphysics of pragmatism, pluralism, and radical empiricism was the result. Thus, his answer was a resounding "yes!" except that the brand of reductionistic positivism that passed for scientific psychology would itself have to undergo a radical transformation in the process.

James's Tripartite Metaphysics

It is my contention that James's tripartite metaphysics of pragmatism, pluralism, and radical empiricism are most clearly articulated in his *Varieties of Religious Experience* (1902), at least to the extent that he had been able to develop these constructs up to that time.[10] Pragmatism was a way of evaluating the manner in which a belief meant something to the individual—the so-called cash value of an idea. It also became a method for evaluating truth claims about the nature of ultimate reality. Though they might seem at first entirely different, if an evaluation of the efficacy of two competing claims showed they both led to the same ends for the individual holding them, then for all intents and purposes, they were equal. This did not mean that they were the same, however. Rather, the war between which one was true and which one false gave way to a certain ecumenism with regard to tolerating people's differences, while preserving the unique origin of their beliefs (James, 1907).

James's doctrine of pluralism was most clearly expressed in *The Varieties* (1902), particularly in two places; first in his definition of religion, not as a recitation of the creeds of the denominations, but as spiritual experiences that occur within the individual; and second, when he indicated that all human beings were capable of experiencing the highest state of transcendence in which all appeared as One, except that this One may not be the same from person to person. He called it noetic pluralism in his *Pragmatism* (1907) when referring to it as the capacity for visionary knowledge.

Pluralism was the idea that we live in the midst of multiple realities. We do not have just one self, James had proclaimed, we have several. The rational everyday waking state is not the only state of consciousness possible to experience; there are multiple states beyond the margin. There is not just one belief system about the nature of ultimate reality, there are as many conceptions as there are individuals, some banding together in common belief, others not.

But it was radical empiricism that was at the core of his metaphysics. Radical empiricism he defined as pure experience in the immediate moment before the differentiation of subject and object. We might try and conjecture when such experiences are likely to occur. Engineers at MIT have measured in the milliseconds the lag between perception of an object and apperception of the object—the superimposition of all our knowledge about all such objects onto that object. Husserl's phenomenological method of the epoché asks us to suspend judgment about an object, meanwhile contemplating the object as a thing-in-itself, as immediately lived experience, without labels. Complete objectless concentration (*asamprajnata samadhi*) that is described in the Yoga texts (Woods, 1917) and also simulated in the laboratory (Taylor, 1973) would constitute yet other examples, as would the apophatic experience of emptiness or the *via negativa* as James described such experiences in *The Varieties of Religious Experience* (1902).

By the time of *The Varieties*, however, as we have said, pragmatism had soon become an international movement, overshadowing pluralism and radical empiricism, and when philosophers maintained they did not understand radical empiricism, James brushed such questions aside by saying that one did not need radical empiricism to continue the dialogue about pragmatism. I believe that he said this in order not to curb the momentum of this one great idea. What he meant was that he had not yet fully developed radical empiricism, but that he planned to. Meanwhile, let the discussion of pragmatism continue. However, philosophers and psychologists took that to mean radical empiricism and pluralism were irrelevant to pragmatism and could be ignored, a mistake codified since the time of Bertrand Russell up to the present mispronouncements by the late Richard Rorty within the analytic tradition in Western philosophy (Taylor, 2003b).

Despite the success of pragmatism, John McDermott (1977) long ago established that radical empiricism was actually the core of James's metaphysics. James unveiled the concept in 1904 in a pair of articles respectively entitled "Does Consciousness Exist?" and "A World of Pure Experience" (James, 1904a, 1904b). To the question "Does Consciousness Exist?" James scandalized the psychologists and philosophers by declaring "no!" It did not exist as an independent object of study, but it does always exist as a function of someone's personal consciousness somewhere. It was a topic extensively written about in the professional literature of the time, but it was always hypostatized by the scientist as an independent object, when in fact it was a process. It was as if the scientists and philosophers themselves were not subject to its vagaries, because they were the ones studying it. Their position, based on the logical ordering of sense data alone, which they still consider the only route to truth, led them to believe that because they were hyper-objective, they believed they were untainted by subjectivity, and allegedly had no underlying metaphysical assumptions, leading them somewhat narcissistically to the conclusion that they were the only human beings on the face of the earth who were in possession of what is called reality.

Points of Comparison

From James's standpoint, we may divine several points of common comparison between James and the modern neurophenomenologists:

James was, first of all, anti-reductionistic. If the 1878 Lowell Lectures tell us anything, it is that James was more like a systems thinker of today. In the laboratory, he worked as an atomist (James on Hall; Taylor on James), focusing in on specific areas of brain function when conducting experiments on brain neurophysiology. Meanwhile, he and his colleagues at Harvard Medical School, such as Henry Pickering Bowditch and James Jackson Putnam, represented what Bowditch had called a "synthetic physiology," which in the hands of James's students, such as Walter Bradford Cannon, continued long into the twentieth century as a "whole-organism physiology." But at the psychological level, James clearly argued for the full spectrum of experience, including dynamic theories of the subconscious or the subliminal. Moreover, he argued in *The Varieties of Religious Experience*, in his first chapter on "Religion and Neurology," that spiritual experience could not be reduced merely to either the language of physiology, or the diagnosis of psychopathology.

Compare this with Varela et al. (1991) enactive view, in which consciousness is a distributed phenomenon over the whole active organism, not just the brain embedded in its environment. As a remedy for the "Hard Problem," the neurophenomenologists do not seek to oppose the subject to the object, but move beyond this split to their fundamental correlation through awareness of reciprocal constraints and codeterminism (Lutz & Thompson, 2003; Varela, 1996).

Second, James advocated the overthrow of representationalism. Radical empiricism rejects the idea that there are always two of the same object—the one out there in the material world and the other that is modeling that object in the mind of the observer. Rather, radical empiricism posits that there is only one object and that exists at the intersection between the history of that particular object in the immediate moment and the autobiography of the observer who is there perceiving it. Zahavi and Parnas (1999) give a slightly different but similar critique of the theory of representation in terms of the constructs of current neuroscience.

Third, both James and the neurophenomenologists argue that the subjective and the objective both operate legitimately within the same larger theater of experience. James called this knowledge about a thing versus acquaintance with it. Meanwhile, the neurophenomenologists take this as the very heart of the Hard Problem, which is the search for the solution to the relation between the brain and the mind, between direct experience of a thing and our scientific models of it.

Fourth, James pointed out the fallacy of hyper-objectivism. The belief that all subjective factors could be eliminated from the experimental situation while the experimenter took the position of objective neutrality was a mythic ideal that is never reached in any experiment, because James said that regardless of the claims of the reductionistic empiricists about their own purity, "the juices of metaphysical assumptions leak in at every joint." In chapter 7, "The Methods and Snares of

Psychology," in *The Principles* he called this "The Psychologist's Fallacy," the tendency of the so-called objective scientist to confuse his own state of mind for the state of mind of the subject whom he is studying (James, 1890, v. 1, p. 196).

Fifth, James argued for what we would today call an enactive or embodied approach to understanding the relation between the organism and the environment. In "A World of Pure Experience" (1904a), he said that our actual experience of consciousness probably resides in our breathing. Our experience of the object at the center of the waking conscious field we know belongs to us because thoughts are always warmed by our emotions. This is how we know they belong to us. The interesting problem is that the stream of thought is always dominant, while the stream of feeling often flows onward just beyond the margin of waking consciousness. Our physiology is always there behind every cognition, close by. This, of course, is the neurophenomenologists' prime agenda—to widen the purview of the cognitivist away from an all-to-narrow focus on behavior, from which the cognitive movement in psychology sprang, and toward the context that a thought always occurs in a physical body that is situated in a specific environment.

Sixth, both James and the neurophenomenologists argue for the relation of subject and object as one of intersubjectivity. Subject and object do not mutually exclude each other, but are interfused, where each defines the other. Compare, for instance, with Zahavi (2001).

Finally, it should come as no surprise as the Jamesean tradition made inroads into the early development of phenomenology even before Husserl (Taylor, 2012b; Tymienieka, 1989). There is also historical evidence that James's influence on Husserl came through Carl Stumpf, who first introduced Husserl to James's *Principles of Psychology* (1890). Even most Husserlians themselves do not know, however, that Husserl was introduced to James's later radical empiricism just after he completed his own *Logic* (Husserl, 1900–1901). In the Husserl Archives at Louvain, for instance, there is a copy of James's "A World of Pure Experience" (1904a), signed "from the Author" (Taylor). We may contrast this with Varela's own interest in the early days of phenomenology.

What the Neurophenomenologists Have Been Saying About James

Meanwhile, we may penetrate a little further into this line of thinking by also asking what parts of James the neurophenomenologists have been referring to in their own work. We shall only take a representative example here rather than an exhaustive one because the nuances have occasionally been extraordinarily brief; though in other places have gone into significant detail.

Varela (1999), for instance, has focused on "the specious present" and applied it to the neurophenomenology of time consciousness, according to Gordon (2009). Moreover, according to Varela, himself, his title from his reading of William James's discussion of "the specious present" in *The Principles of Psychology* (1890, vol. 1) is not a comparison of James and Husserl, but more a rendition of Varela's neurophenomenological sense of time consciousness.

In this article, Varela (1999) engages in a detailed analysis of James's own discussion of the topic:

> We are constantly conscious of a certain duration—the specious present—varying in length from a few seconds to probably not more than a minute, and that this duration (with its content perceived as having one part earlier and the other part later) is the original intuition of time. Longer times are conceived by adding, shorter ones by dividing, portions of this vaguely bounded unit, and are habitually thought by us symbolically. Kant's notion of an *intuition* of objective time as an infinite necessary continuum has nothing to support it. The *cause* of the intuition which we really have cannot be the duration of our brain-processes or our mental changes. That duration is rather the *object* of the intuition which being realized at every moment of such duration, must be due to a permanently present cause. This cause—probably the simultaneous presence of brain-processes of different phase—fluctuates; and hence a certain range of variation in the amount of the intuition, and in its subdivisibility accrues. (p. 642)

As did James, Varela (1999) used this construct to illustrate how "lived time" is not physical-computational, but existential-phenomenological. James expressed "awareness of change [as] the condition on which our perception of time's flow depends" (1890, p. 620). He posited a relationship between the bare phenomenon or the immediately known thing and the phenomenon of changing brain states that is cognizant of its object. James argued that the entire brain process *is* the state of consciousness, the soul a medium upon which these processes combine their effects. However, for James, and Gordon also would argue for Varela as well, the psychophysical parallelism within the dualism of thought and actuality, mind and matter, were both aspects of, or structures formed from, a more fundamental stuff—pure experience, which was already reified into preestablished categories by the senses before psychology began its work.

Gallagher (1998) has also taken up a similar discussion in *The Inordinance of Time*. There, Gallagher attempted to reconcile the temporality of experience in cognitive science with the challenges of poststructuralism (p. 4) at the point where they intersect with phenomenology. He declared Husserl the voice who solves the problem that James sets up in his discussion of the specious present, namely, seeds of the past and the future must exist simultaneously in our experience of the present moment, though they are themselves not yet full-fledged memories nor projections of what is to come. Gallagher refers to a corrective of this view as Husserl's "cognitive paradox" where:

> (1) The succession must be perceived as a momentary, individual, and durationless act of consciousness, and (2) the perception of succession requires continuity of both sensations and memory images. Thus, succession can only be understood in terms of a momentary simultaneity in the present. (p. 4)

In my view, James's references to radical empiricism could be productively considered here. James refers to the stream of consciousness in his *Exceptional Mental States Lectures* (Taylor, 2010a) as a field with a focus and a margin, where the margin keeps changing as the stream flows onward. We have the monoideism of the trance state in hypnosis and the *idea fixe* in psychopathic compulsions, all the way at the other end of the spectrum to the limitless expansion of consciousness in the apophatic experience of the transcendent mystical state. Both Husserl and Gallagher limit themselves to what James says about time consciousness in *The Principles* (1890) in the present argument, however. And it is also true that James did not formally coin the phrase "radical empiricism" until 1897. In any case, Gallagher does enter into a discussion of the expansion and contraction of consciousness around the immediate moment as a factor in the specious present. Gallagher himself has said:

> I'm a big fan of James from my graduate school days when I studied pragmatism with the late Isabel Stearns. I agree there are some interesting connections between radical empiricism and phenomenology, and also James's notion of the specious present and Husserl's time-consciousness (Personal communication, Feb. 20, 2008).

As recently as 2005, Zahavi (2005) also invokes James with regard to the Husserlian idea of time consciousness (p. 67).

Gallagher (2005) also invokes James elsewhere. He takes up the distinction between consciousness and embodiment and poses that body image and body schema arise out of what James had called the "blooming, buzzing confusion" (p. 65). Gallagher's contention was that such perception, with regard to first perceptions of the infant, did not arise until 3–6 months after birth, which has long been the standard assumption, but he contends there is new evidence challenging that interpretation, but still based on the idea that perception arises out of experience. Normative psychologists still believe the sensory systems remain separate from each other, when Merleau-Ponty took a more phenomenological point of view that Gallagher referred to as multimodal.

In my view, the argument against the developmental psychologists is that many still think that perception means seeing with the physically intact structures of vision, and if there is no vision, there is no perception. James put no specific time limit on when "the blooming buzzing confusion" emerges as a perception in the infant, but he did say that in the evolutionary scheme of things, all attention was based on interest and that the true wonder of perception is not judged solely on what we see but also on what we do not see. A perception must emerge out of the blooming, buzzing, confusion for us to focus our attention on the object, but once done, all that we are not interested in is kept at bay. Perception therefore is based on what we see and, as well, all that we do not see. James did say that perceptions are made over long periods of time and that "Every perception is an acquired perception" (Gallagher, 2005, p. 157). However, Gallagher did review new evidence to suggest that perception based on body schema exists from birth, that sensory systems are already multimodal, and that one sense modality does educate other sensory modalities.

My own conclusion is that while the neurophenomenologists were familiar with James's radical empiricism, their primary focus has been on James's *Principles of Psychology*, not his radical empiricism of 1904. One gets the sense on reading through Varela that radical empiricism was on the tip of his tongue. Had he known the details of James's thought on that subject, as well as he and his colleagues have mastered *The Principles*, neurophenomenology might have broken through already in their task to introduce phenomenology to neuroscience.

Implications for Psychology

To understand the implications of neurophenomenology for psychologists, in the first instance, it is important to make a general comment about the relation of the neurosciences to reductionistic empiricism in mainstream psychology. While there were certainly other forces at work from the 1930s onward pushing psychology more toward social, cognitive, and developmental thinking, the core definition of experimental psychology remained learning theory, mathematical psychology, sensation and perception, animal rat motivation, the comparative study of lower order primates, as well as physiological psychology. Personality, abnormal, social and clinical psychology were considered the soft scientific underbelly of the discipline, the critics of which tried mightily to maintain psychology as an experimental science despite these fields.

The major thrust of the era academically was to so-called Age of Theory, where the attempt was made to subsume all research under Clark Hull's hypothetico-deductive method, with an eye toward psychology's definitive contribution to the Unity of Science Movement within the physical sciences. World War II, among other forces, disrupted and eventually dissipated psychology's efforts in that direction. The result after the war was an abandonment of the search for a single grand theory for all of psychology and an opening of a new era of competing micro-behavioral theories of the organism. Behaviorism had been routed, except that the discipline was still controlled by the experimentalists, who were taking over the experimental laboratories from the rat runners and putting themselves forward as cognitive psychologists. This is why the era of cognitive psychology that followed is still often referred to as cognitive-behaviorism. Cognitive psychology was birthed out of the lineage of behaviorism, not physiology.

Clinical psychology, meanwhile, had become more psychodynamic, and more psychosomatically oriented, by the 1960s more existentially and humanistically oriented, and thus more of a challenge to the hegemony of the experimentalists. Sigmund Koch (1958–1964), America's foremost expert in the application of logical positivism to positivistic empiricism, conducted a survey for the American Psychological Association at mid-century on the status of measurement to theory. In six volumes, the work entitled *Psychology: The Study of a Science*, and in league with 89 of the highest-ranking experimental psychologists in the West, concluded that the correlation was so low as to be embarrassing. At the same time, the debates between Carl Rogers and B. F. Skinner that prevailed in the late 1950s and early 1960s had pitted behaviorism against phenomenology when phenomenological and existential psychology were on the ascendant. Though they had finally been incorporated into the American Psychological Association, clinicians soon outpaced the experimentalists in numbers and eventually took control of the APA.

Meanwhile, physics and biology began talking to each other in the 1950s and by the 1970s had spawned the neuroscience revolution, which was essentially about the biology of consciousness. Neuroscientists swept down into all the traditional scientific disciplines and plundered whatever they found there regarding the brain, which left cognitive psychologists with their counterparts in behavior, while at the same time creating a new subdiscipline called cognitive neuroscience, an endeavor still not fully understood by the cognitivists who remained.

So we have three levels of science to remember, cognitive psychology, cognitive neuroscience, and neuroscience proper. The neurophenomenologists are clearly speaking from the cognitive neurosciences to the larger field of neuroscience. However, neurophenomenology has yet to reach the radar screen of cognitive psychologists and psychology in general. It is also my contention that the neurophenomenologists have captured recent trends in the neurosciences in neurophilosophy, the philosophy of mind, and what d'Aquili and Newberg have called neurotheology, by taking a first step engaging the neuroscientists proper, not with more analytic philosophy, but with phenomenology as a formal discipline.

Phenomenology had entered psychology through the existential-phenomenological movement in the 1950s, spearheaded by psychotherapists in psychology and psychiatry around such authors as Rollo May (May, Angel, & Ellenberger, 1958), Victor Frankl in *From Death Camp to Existentialism* (1959), and Humanistic psychologists in league with May, such as Gordon Allport, Carl Rogers, and Abraham Maslow (May, 1961), and others around Clark Moustakas (1961) and Amedeo Giorgi (1970).

Giorgi, however, is not a clinician, but a trained experimental scientist who sees himself clearly as a phenomenological psychologist. He has been most profoundly influenced by Husserl and Merleau-Ponty in the tradition of the Continental phenomenologists, but has also written on William James, and at one point translated Linschoten on radical empiricism and phenomenology. Giorgi's position is that phenomenology is, in fact, the unifying element in psychology regarding the war between the experimentalists and the clinicians. A phenomenologically oriented psychology would be able to explore the commonalities between patient and therapist as well as experimenter and subject and would also move psychology more toward becoming a descriptive science rather than the current approach, which is one devoted to the manipulation, prediction, and control of the subject while the experimenter adopts an essentially false position of neutrality in the conduct of the study.

James's radical empiricism is in complete agreement with Giorgi, but only up to a point, which involves the differentiation of phenomenology in science from its role in psychology. Such an important distinction James himself dealt with in 1904 in correspondence with the logician and philosopher of science, Charles Sanders Peirce [pronounced "purse"]. James had sent his article on "Does Consciousness Exist?" to Peirce and the response he got was largely unintelligible, though one thing was clear. Peirce maintained that phenomenology was absolutely fundamental to science, but James was in error calling it psychology. The issue appeared to be where Peirce saw psychology in the hierarchy of the sciences. It was a formal discipline, but a branch of what would become the social sciences, as most psychologists interpret it today.

James, on the other hand, argued for a new kind of psychological science, based on an intuitively oriented, phenomenological psychology in the immediate moment (Taylor, 2010a, b, c). In this, he rejected the artificial boundaries between the disciplines as man-made and thus subject to the same vagaries as all rational thought based on an analysis of the senses alone. The distinctions were certainly rational, but rationality is a sentiment whose reign continues by consensus alone, precluding other ways of knowing. The importation of the Western scientific model globally would be a case in point (see, for instance, *The Oxford Handbook in the History of Psychology*, Baker, 2012). It can only be considered the New Colonialism to presume that reductionistic empiricism fields the only definition of reality possible. The consequence is that the globalization of reductionistic materialism in psychological science precludes research psychologists from ever really listening to the indigenous psychologies of non-Western cultures, which have their own cultural mythologies, standards of mental health, and unique models for the actualization of human potential for persons within their own cultures.

Giorgi maintains a more traditional view of psychology, while James maintained that there is no science anywhere without persons—that psychology is epistemology. The implication is that a phenomenologically oriented psychology of persons is foundational to all the basic sciences, not physics. Giorgi, while adhering to the traditional distinction between the Natural and Human Sciences, believes that James's position traditional phenomenologists would label psychologism (personal communication, March 25, 2010). This is tantamount to taking a branch of science and calling it the trunk and roots, at least according to the traditional and current view of psychology. But phenomenology now exists in a new world, where it implicates the experimenter in the conduct of science, which was James's point to begin with.

We may count James's functionalism as the first order expression of his radical empiricism. The lineage was carried on not by the New Realists, but by the macropersonality theorists of the 1930s and 1940s in the work of Gordon Allport, Henry Murray, and the Murphys, who, through the writings of Allport, represented a science of the whole person (Taylor, 1992). While their students endeavored to integrate themselves into psychology as a social science of the times, instead, these pioneer theorists themselves became godfathers and godmothers of the Humanistic Movement in psychology that sprouted in the 1940s and 1950s (Taylor, 2000; Taylor, Martinez, & Martin, 2000). The Humanistic Movement, in turn, argued for psychology as a person-centered science, which was not Humanism and advocated not just a psychology of persons, but a much more sophisticated and nuanced epistemological critique of existing psychological systems. Humanistic psychology as a global movement was absorbed almost completely into the Human Potential Movement after 1969, however, and attempts to focus on a psychology of the person were displaced by the cognitive revolution as behaviorists bought themselves computers in droves and developed a more mentalist language of behavioristic principles. Cognitive information processing models, computational analysis, and artificial intelligence on the computer screens soon replaced kibbles in the dish of the white rat, while any vision of the whole person faded before the ascendency of the Five Factor Model. The person seemed to have disappeared as a focus in psychology.

The humanistic implications of the neuroscience revolution, however, namely, the return of the philosophical questions once banned from discussion during the cognitive-behavioral era, implicate the very personality of the experimenter in the conduct of psychological science now that the revolution in consciousness is forcing science to study its own methods and points of view. I refer to this as the "phenomenology of the science making process itself" and see it reflected in the work of such investigators as Max Velmans, University of London, who maintains that the experimenter represents the new variable that must be accounted for in the puzzle to understand the relation between the brain and the mind.

Neurophenomenologists carry this argument one step forward by challenging neuroscience to train the next generation of scientists in the methods of phenomenology. This alone is destined to alter the way the scientific method is taught in experimental psychology. However, in all likelihood, it must play itself out in psychology as science generally before experimental psychologists are prepared to handle the shift from a nineteenth-century definition of the physical sciences in which their kind of science, based on a Newtonian, Kantian, and Aristotelian model, has prevailed for the past 100 years to a more experiential, phenomenological, and Jamesean view of psychology as a person-centered science.

Notes

1. Five core ideas that define the enactive paradigm are the mutually supporting concepts of *autonomy*, *sense-making*, *embodiment*, *emergence*, and *experience* (Di Paolo, Rohde, & De Jaegher, 2007; Thompson, 2005, 2007; Varela et al., 1991). A key principle of the enactive approach is that the organism is a center of activity in the world. See also *Phenomenology and the Cognitive Sciences*, *Participatory Sense-Making: An Enactive Approach to Social Cognition* (De Jaegher & Di Paolo, 2007).
2. The lectures on "Exceptional Mental States" James never published (Taylor, 1982/2010a), but their content was bequeathed to the next generation of James's graduate students, such as Boris Sidis (1898). The content of the 1896 lectures also found their way into several major chapters in James's *Varieties of Religious Experience* (1902). James made the case there that the road to ultimately transforming states of mystical consciousness lay through an exploration of dynamic subconscious or subliminal states within the individual.
3. Francis J. Child, who had been in attendance, had written to James Russell Lowell that the subject of the lectures was the brain as the organ of the mind, though I do not think James would completely agree (James, 1988, p. 449). James's later radical empiricism suggests that the brain as the organ of the mind would be just as appropriate. Each point of view is true with regard to a different level of experience.

   Biological reductionists would claim superiority. Radical empiricists would accept both as true because both are based on belief—that is, both are based on unproven assumptions about the nature of ultimate reality.
4. The origin myth that Wundt founded the first laboratory in Leipzig in 1879, I have dealt with in Taylor (2003a). The Hypothesis of the Three Streams: Or, Why Wundt's Laboratory was not founded until 1950.
5. See, for instance, James (1874).
6. This, of course, was 15 years before Ramon y Cajal won the Nobel Prize for identifying the neuron.
7. Marie Jean Pierre Flourens (13 April 1794–6 December 1867), father of Gustave Flourens, was a French physiologist, a founder of experimental brain science, and a pioneer in anesthesia. Eduard *Hitzig* (February 6, 1839–August 20, 1907) was a German neurologist and neuropsychiatrist. Sir David Ferrier, FRS (13 January 1843–19 March 1928) was a Scottish neurologist and psychologist.

8. See, for instance, James (1875). While phrenology might not seem an apt scientific comparison, its early pioneer, Franz Joseph Gall, sparked a vigorous discussion among the early brain neurophysiologists abroad in the controversy over localization of function in brain neuroanatomy.
9. James (1879b), a year after the Lowell Lectures, wrote on the Sentiment of Rationality, more fully developed in 1897 in *The Will to Believe and other Essays in Popular Philosophy.*
10. Moreover, as I have argued elsewhere, his tripartite metaphysics was a retailing of Peirce's Three Categories of Firstness, Secondness, and Thirdness, but in a more typical Jamesean vein. See Taylor (2012b).

References

Baker, D. B. (Ed.). (2012). *The Oxford handbook of the history of psychology: Global perspectives*. Oxford, NY: Oxford University Press.

De Jaegher, H., & Di Paolo, E. A. (2007). Participatory sense-making: An enactive approach to social cognition. *Phenomenology and the Cognitive Sciences, 6*(4), 485–507.

Di Paolo, E. A., Rohde, M., & Jaegher, H. (2007). Horizons for the enactive mind: Values, social interaction, and play. In J. Stewart, O. Gapenne, & E. A. Paolo (Eds.), *Enaction: Towards a new paradigm for cognitive science*. Cambridge, MA: The MIT Press.

Frankl, V. E. (1959). *From death-camp to existentialism: A psychiatrist's path to a new therapy*. Boston, MA: Beacon Press.

Gallagher, S. (1998). *The inordinance of time*. Evanston, Ill: Northwestern University Press.

Gallagher, S. (2005). *How the body shapes the mind*. New York, NY: Oxford University Press.

Giorgi, A. (1970). *Psychology as a human science: A phenomenologically based approach*. New York, NY: Harper & Row.

Gordon, S. (2009b). Radical empiricism and the neurophenomenology of time. Unpublished working paper.

Husserl, E. (1900–1901). *Logische Untersuchungen. Zweiter Teil: Untersuchungen zur Phänomenologie und Theorie der Erkenntnis*. [*Logical investigations*. 2 vols.: *Studies in phenomenology*; *The theory of knowledge*.] Halle a.S.: Max Niemeyer.

James, W. (1865). Review of Alfred R. Wallace's The origin of human races and the antiquity of man deduced from the theory of 'natural selection.' *North American Review, 100*, 290–298.

James, W. (1868). Review (unsigned) of Charles Darwin's The variation of animals and plants under domestication. *Atlantic Monthly, 22*, 122–124.

James, W. (1874). Notice (unsigned) of J. B. Petigrew's Physiology of the circulation in plants, in lower animals, and in man. *Nation, 19*, 63.

James, W. (1875). Notice (unsigned) of F. Papillon's Nature and life. *Nation, 20*, 429.

James, W. (1876). Note (unsigned) on the *Revue Philosophique*. *Nation*, *22*, 147.

James, W. (1877). Review (unsigned) on the Mind and the Brain (D Ferrier, H. Maudsley, J. Luys). *Nation, 2J h*, 355–356.

James, W. (1879a). Are we automata? *Mind, 4*, 1–22.

James, W. (1879b). The sentiment of rationality. *Mind, 4*, 317–346.

James, W. (1890). *The principles of psychology* (2 vols.). New York, NY: Henry Holt.

James, W. (1892). *Psychology: Briefer course*. New York, NY: Henry Holt.

James, W. (1895). The tigers in India. *Psychological Review*, *2*, 105.

James, W. (1897). *The will to believe and other essays in popular philosophy*. New York, NY: Longmans, Green & Co.

James, W. (1898, September). Philosophical conceptualizations and practical results. *University of California Chronicle*, *1*(4), 24.

James, W. (1902). *The varieties of religious experience: A study in human nature*. New York, NY: Longmans, Green and Co.

James, W. (1904a). A world of pure experience. *Journal of Philosophy, Psychology and Scientific Methods, 1*(21), 533–543, 561–570.

James, W. (1904b). Does "consciousness" exist*? Journal of Philosophy, Psychology and Scientific Methods, 1*(18), 477–491.

James, W. (1907). *Pragmatism*. New York, NY: Longmans, Green, and Co.

James, W. (1988). The 1878 Lowell Lectures. In F. Burkhardt, F. Bowers, & I. K. Skrupskeis (Eds.), *Manuscript lectures*. *The works of William James*. Cambridge, MA: Harvard University Press.

Koch, S. (Ed). (1958–1964). *Psychology: The study of a science* (7 vols.). New York, NY: McGraw Hill.

Lutz, A., & Thompson, E. (2003). Neurophenomenology: Integrating subjective experience and brain dynamics in the neuroscience of consciousness. *Journal of consciousness Studies, 10*(9–10), 31–52.

Maturana, H. R., & Varela, F. J. (1987). *The tree of knowledge: The biological roots of human understanding*. Boston, MA: New Science Library.

May, R. (Ed.). (1961). *Existential psychology*. New York, NY: Random House.

May, R., Angel, E., & Ellenberger, H. F. (Eds.). (1958). *Existence: A new dimension in psychiatry and psychology*. New York, NY: Basic Books.

McDermott J. J. (Ed.). (1977). *Writings of William James: A comprehensive edition*. Chicago, IL: University of Chicago Press.

Moustakas, C. E. (1961). *Loneliness*. Englewood Cliffs, NJ: Prentice-Hall.

Sidis, B. (1898). *The psychology of suggestion*. New York, NY: Appleton and Co.

Taves, A. (1999). *Fits, trances, and visions: Experiencing religion and explaining experience from Wesley to James*. Princeton, NJ: Princeton University Press.

Taylor, E. I. (1973). *Psychological suspended animation: Heart rate, blood pressure, time estimation, and introspective reports from an anechoic environment*. Unpublished MA Thesis, Southern Methodist University, Dallas, TX.

Taylor, E. I. (1990). William James on Darwin: An evolutionary theory of consciousness. *Annals of the New York Academy of Sciences*, *602*, 7–33. Originally presented as part of the Darwin Centennial Lectures.

Taylor, E. I. (1992b). The case for a uniquely American Jamesian tradition in psychology. In M. Donnelly (Ed.), *Reinterpreting the legacy of William James* (APA Centennial William James Lectures). (pp. 3–28).Washington, D.C.: American Psychological Association.

Taylor, E. I. (2000). "What is man, Psychologist, that thou art so unmindful of him?": Henry A, Murray on the historical relation between classical personality theory and humanistic psychology. *Journal of Humanistic Psychology*, *40*(3), 29–42.

Taylor, E. I. (2003a). *Why Wundt's laboratory was not founded until 1950, Or the hypothesis of the three streams*. The Wallace Russell Memorial Lecture presented to (Division 26) The History of Psychology at the 109th annual meeting of the American Psychological Association, Toronto, Canada, August 9, 2003.

Taylor, E. I. (2003b). Have we engaged in a colossal misreading of James's *Varieties*? *Streams of William James*. Newsletter/journal of the William James Society.

Taylor, E. I. (2010a). *William James on exceptional mental states: Reconstruction of the 1896 unpublished Lowell lectures* (2nd printing). Portsmouth, NH: Jetty Press. (Original work published 1982)

Taylor, E. I. (2010b). William James on an intuitive, phenomenologically oriented psychology in the immediate moment: The true foundation for a science of consciousness? Special issue on Consciousness in the Neurosciences. *History of the Human Sciences, 23*(3), 119–130.

Taylor, E. I. (2010c). William James and the humanistic implications of the neuroscience revolution: An outrageous hypothesis. *Journal of Humanistic Psychology, 50*(4), 410–429.

Taylor, E. I. (2012a). The ontological triad in James and Peirce: A moment in The history of consciousness, *Revue Internationale de Philosophie, 260*, 177–186.

Taylor, E. I. (2012b). The stream of consciousness: Literary psychology as the first uniquely American phenomenology in the works of William James and his Swedenborgian and

Transcendentalist milieu. In A.T. Tymieniecka (Ed.), *Art, Literature and Passions of the Skies, Analecta Husserliana, 108*, 112.

Taylor, E. I., Martinez, S., & Martin, F. (Eds.). (2000). *Old Saybrook One: Benchmark in the history of humanistic psychology. A commemorative edition*. Cambridge, MA: The Essene Press.

Thompson, E. (2005). Sensorimotor subjectivity and the enactive approach to experience. *Phenomenology and the Cognitive Sciences, 4*(4), 407–427.

Thompson, E. (2007). *Mind in life: Biology, phenomenology, and the sciences of mind*. Cambridge, MA: Harvard University Press.

Tymienieka, A.T. (1989). The theme: The history of American phenomenology-in-process. In E. F. Kaelin & C. O. Schrag (Eds.), *American phenomenology: Origins and development, Analecta Hussereliana, 26*, (ix-xxvii). Brattleboro, VT: World Institute for Advanced Phenomenological Research and Learning.

Varela, F. J. (1996). Neurophenomenology. A methodological remedy for the hard problem. *Journal of Consciousness studies, 3*(4), 330–349.

Varela, F. J. (1999). The specious present. A neurophenomenology of time consciousness. In J. Petitot, F. J. Varela., B. Pachoud, & J-M. Roy (Eds.), *Naturalizing phenomenology: Issues in Contemporary Phenomenology and Cognitive Science* (pp. 266–314). Stanford, CA: Stanford University Press.

Varela, F. J., & Shear, J. (Eds.). (1999b). *The view from within: First-person approaches to the study of consciousness*. Thorverton, UK: Imprint Academic.

Varela, F. J., Thompson, E., & Rosch, E. (1991). *The embodied mind: Cognitive science and human experience*. Cambridge, MA: The MIT Press.

Weeks, E. (1966). *The Lowells and their institute*. Boston, MA: Little, Brown and Co.

Woods, J. H. (Trans.) (1917). *The yoga sutras of Patanjali. Harvard Oriental Series, 17*, Cambridge, MA: Harvard University Press.

Zahavi, D. (2001). Beyond empathy: Phenomenological approaches to intersubjectivity, *Journal of Consciousness studies, 8*(5–7), 151–167.

Zahavi, D. (2005). *Subjectivity and selfhood*. Cambridge, MA: Bradford/The MIT Press.

Zahavi, D., & Parnas, J. (1999). Phenomenal consciousness and self-awareness: A phenomenological critique of representational theory. In S. Gallagher & L. Shear (Eds.), *Models of the self* (pp. 253–270). Exeter, UK: Imprint Academic.

Psychoneurointracrinology: The Embodied Self

Susan Gordon

> *The spiritual or value life falls well within the realm of nature, rather than being a different and opposed realm. It is susceptible to investigation at once by psychologists and social scientists and will eventually become also a problem for neurology, endocrinology, genetics, and biochemistry as these sciences develop suitable methods.*
>
> (Maslow, 1971, p. 320)

Is there an embodied, growth-oriented dimension of the person[1] or *neurophenomenological self* that is biochemically and energetically rooted in the hypothalamic-pituitary-gonadal (HPG) and hypothalamic-pituitary-adrenal (HPA) axes of the body? Neurophenomenology encourages a vision of the integrated self through a reexamination of the relationship between the subjectively experienced mind and the objectively known material substratum of the mind, the brain. Enactive cognition[2] points out that while it is impossible to infer such engagement through the brain, it is possible through the lived experience of the body.[3] *Psychoneurointracrinology* explores this explanatory gap between the mind and the brain. This construct represents the interrelationship between psychological, neurological, and intracrinological processes forming a mind-brain continuum within the person (Gordon, 2001, 2007, in press a, b). Psycho (psychological) refers to constructs variously referred to as psyche, self, soul, mind, and consciousness. Neuro (neurological) refers to the composition and reactions within the nervous system. Intracrine (intracrinological) refers to the intracellular biosynthesis of steroids: the binding of receptors and the formation of enzymes that catalyze the creation of hormones within the cell.

S. Gordon, Ph.D. (✉)
Department of Psychology, National University, La Jolla, CA, USA

Southbury Clinic for Traditional Medicines, 900 Main Street South,
Southbury, CT 06488, USA
e-mail: susan.gordon@snet.net

S. Gordon (Ed.), *Neurophenomenology and Its Applications to Psychology*,
DOI 10.1007/978-1-4614-7239-1_5,

It is argued that the HPA-HPG axes have neural correlates, which are responsible for enactive engagement and the development of meaning through their connections to the higher-order functions of the brain revealing the person's sense of self and well-being. Two theories of enactive cognition are advanced: the theory of psychoneurointracrine autopoiesis, which examines how the regulation of a steroid's receptor is modulated by the person's perception of experience, and the theory of emergent global states, which explains how corticolimbic projections of the HPG-HPA axes integrate prereflective, autonomic, and subliminal experience in the development of meaning and emergence of self.

Psychoneurointracrinology, as a complex system,[4] differs from psycho-neuroimmunology and psychoneuroendocrinology. Each of these systems takes an interdisciplinary approach to research and clinical practice; however, psychoneuroimmunology and psychoneuroendocrinology focus on the diagnosis and treatment of stress-related disease,[5] and psychoneurointracrinology focuses on health, holism, and self-actualization. Endocrinology refers to the macro-level branch of medicine that examines ductless glands (pituitary, pineal, thyroid, thymus, adrenal, pancreas, ovary, testes), which secrete hormones directly into the bloodstream. Its subfield, intracrinology,[6] refers to the intracellular, micro-level synthesis and regulation of peptides, hormones,[7] and their enzymes that bind to peripheral receptor sites where steroids allosterically unlock the expression of the receptor's function or activate triggering compounds, within the cytoplasm of cells containing receptors.

Multiple adaptive metabolic and psychological influences modulate the subtle relationship between hormones. As the product of the conservation of autopoiesis and adaptation, states of the brain transform neural networks and states of the nervous system transform the state of the organism, through the structural coupling of sensory and motor surfaces. Neurointracrine systems self-create through propagating waves of chemical concentrations, receptor modulation, and neuronal signaling, which present a complex pattern of spatiotemporal organization that defies the linear concept of feedback. Consciousness (the psychological) embodied as felt sense or consciousness of self is a multidimensional subjective stream of thoughts, perceptions, imaginations, representations, memories, and emotions that are associated with a global state among neuronal networks. These networks are continuously shaped by the person's perception of experience, sense of self and well-being, or *existential-phenomenological state*.[8] Self is not a static object of representation, but a dialectical process of continual growth and change.

Flow of the Argument

I begin with a discussion of autopoiesis and the sense of self with focus on Varela's concept of self and no self. Following this is a discussion of prereflectivity, the embodied self, and the neurophenomenology of emotion. Empirical research shows how the amygdala, a prime mover of the CRF-ACTH pathway (along with the hippocampus and prefrontal cortex), is responsible for the state-dependent perception of the observer. This is followed by a discussion of the growth-oriented dimension of the person and the neurophenomenological self. Next I present a review of theoretical and empirical literature on the human menstrual cycle to advance my theories of psychoneurointracrine autopoiesis and emergent global states. In conclusion, I discuss implications of the psychoneurointracrine model for the study of the person.

Autopoiesis and the Sense of Self

In the early 1970s, biologists argued that the living cell, when conceptualized as an autopoietic or self-producing system, is the continual creation of itself. Varela (1979) defined the unity of autopoietic systems as organized networks of the processes of transformation and destruction through which the system continuously regenerated and realized the processes or relations that produced it. Maturana and Varela (1987), Varela (1979), and Maturana, Varela, and Uribe (1974) adapted principles from cybernetics[9] and dynamic systems theory to explain how structural change within a biological system defines its unity, identity, stability, and internal coherence. Autopoiesis sparked a new way of thinking about the structural determinism of biological systems and the roots of human understanding.[10]

The theory of autopoiesis revisited and called into question theories regarding the unity of consciousness, atomism, empiricism, and representationalism, of great concern to philosophers of the classical modern era such as Descartes, Leibniz, Kant, Hume, Brentano, James, and Peirce. For example, the question for Kant (1781/2003) was determining the form that cognition had to take to make the experience of the world possible. He found pure knowledge in universal a priori, theoretical and deductive, not empirical categories, and claimed that only through representation could we know something as an object (phenomena) given that the law of causality, founded on mathematics, has its a priori basis in human understanding. In contrast, James (1890, 1892/1961) believed that a thing could not be known through its representation, but must be directly experienced. However, the ground itself did not need to be structured before the mind splits pure experience into subject-object and other categories of interpreted experience (James, 1885; Taylor, 1996).[11]

With regard to representation, the central concern of the enactive approach to cognition has remained in contradistinction to the received view that perception is fundamentally the truthful reconstruction of a portion of the physical world through a registering of environmental information. Knowledge is itself enacted. Varela, Thompson, and Rosch (1991) approached the problem of knowledge from a Buddhist perspective arguing that there is no independent, fixed, or unitary self within the world of pure experience. There is, instead, a selflessness or egolessness, a rapidly shifting stream of momentary mental occurrences that include the perceiver and the perceived (in which the Buddhists' five aggregates, constituting a psychophysical complex, made up the person and each moment of their experience).[12] They argued that while self, as an emergent property or process of these aggregates, is empty of self, it is full of experience. Meaning is not located in cognitive symbols or representations, but is instead a function of the global state of the system's complex pattern of activity emerging from the interactions of its many constituents. Varela et al. resolved this problem as follows:

> In effect, if we presuppose the existence of an objective world, independent of us as observers and accessible to our knowledge through our nervous system, we cannot understand how our nervous system functions in its structural dynamics and still produces a representation of this independent world. But, if we do not presuppose an objective world independent of us as observers, it seems we are accepting that everything is relative and anything is possible in the denial of all lawfulness. Thus we confront the problem of understanding how our experience—the praxis of our living—is coupled to a surrounding world, which appears filled with regularities that are at every instant the result of our biological and social histories. (1991, pp. 240–241)

Varela's Selfless Self. Given the enactive view of human knowledge, Varela called for a science of the "sense of self" in which biological cognition was not considered a representation of the world "out there," but rather an ongoing bringing-forth of a world, through the process of living itself. Thus, the theory of autopoiesis attempted to define the uniqueness of the emergence that produced life in its fundamental cellular form, i.e., biochemical pathways of the cell and its membranes continuously regenerate through the internal production of substratum components, and biochemical states of the organism transform the state of activity of neural networks by acting on the neuron's membrane receptors. Having no fixed point of reference, Varela believed that human beings, as autopoietic, homeostatic systems, regenerate, recreating themselves by their own mutual interactions as a natural drift.

Throughout his life, Varela's position remained situated in the context of what he saw as the irreducible nature of conscious experience. He studied phenomenal experience or embodiment as lived from the point of view of the subject's first-person experience associated with cognitive and mental events (i.e., attention, present-time-consciousness, body image, volition, perceptual filling in, fringe, center, and emotion), which represented an irreducible ontological level that retained its quality of immediacy because it played a role in the organism's structural coherence. Consciousness was a distributed phenomenon of the whole active organism, not just the brain embedded in its environment. Rejecting the computational, logical views of the mind in favor of the concrete embodied lived description of its processes, Varela saw the mind as a *selfless* or a virtual self—"a coherent whole that is nowhere to be found, and yet can provide an occasion for the coordinated activity of neural ensembles" (1992, p. 60).

What are the differences between an embodied versus non-embodied approach to cognition? If one uses a non-embodied approach, sensory systems inform the cognitive system and then the motor system does the cognitive system's bidding. The non-embodied view is that there are causal relations between the systems, the sensory and motor systems are not constitutive of cognition. However, if one uses an embodied approach, the relation of the sensorimotor system to cognition is both causally linked and constitutive (Adams, 2010). Human beings as dynamic systems are characterized by a high degree of self-organizing autonomy, and are therefore not reducible to the more basic mental and physical events that constitute them. However, while self may be emergent and constructed, it is not virtual (Mackenzie, 2010, 2012). Self-constitution is an active, embodied, embedded, self-organizing process. Reality is not given. It is instead perceiver-dependent.

As Thompson (2007) notes, the emergent process of self-making is grounded in the fundamentally recursive processes that characterize lived experience: autopoiesis at the biological level, temporalization and self-reference at the level of conscious experience, and conceptual and narrative construction at the level of intersubjectivity. While self may be dependently originated and empty, it is nevertheless real. The Buddhist-enactive conception of the self provides a middle path in which the stream of experience becomes self-referential through the structure of time-consciousness. The embodied being is thus prereflectively aware of itself in and through its active striving body.

The doctrine of nonself or *anātman* that Varela and others refer to is in essence a rejection of the atmān or the enduring substantial self and one of the most controversial aspects of Buddhism. Siderits, Thompson, and Zahavi (2011) offer a wonderful interdisciplinary discussion of consciousness and the self characterized as a type of self-reference, self-illumination, or *sense of self*. Nonself theorists (Krueger, Albahari, Dreyfus, and Ganeri) do not infer from the sense of self that self really exists because they claim that it is not ontologically grounded, but arises from the stream of consciousness. For them, the emergence of self is only "a useful fiction that helps to maintain the sense of agency and mobilize action-guided emotions" (2011, p. 137). In contrast, self theorists (Zahavi, Thompson, Fasching, and Ram Prasad) speak of self as the mode of givenness or essential structure of consciousness, which is prereflective, subjective, and self-revealed (Cai, 2012). They argue that a correspondence between conception and realty is required only if the sense of self is an epistemic state, which takes the self as its objective content. While the self theorists offer no counterargument that defeats the Buddhist doctrine, their characterization of the sense of self (phenomenal character of consciousness) as "self-illuminating" captures the feature, which preconditions the sense of self that the nonself theorists concentrate on.

It is also important to note that Varela derived his concept of "self" and "no self" from immunology as well as Buddhist philosophy. Vaz and Varela (1978) provide an autopoietic framework for understanding the genetic induction and cellular interactions of the immune system. Their hypothesis was that the immune system is a closed network of interactions that self-determines its ongoing pattern of stability and capacities of interaction with its environment. Thus, immune events are a form of self-recognition. For example, in the operation of the lymphoid network, cellular interactions are not only complex, they are self-determined (p. 236). Varela also drew parallels between the nervous and lymphoid systems with regard to ontogenetic development. Every "recognition event" was followed by some form of action that changed the responsiveness of the system through learning that was based on the plasticity of synaptic contacts (p. 245). At variance with the assumptions of the times (e.g., clonal selection theory), Varela and Vaz's picture of the immune system stressed the cooperative nature of events typical of lymphoid cells as a network of interactions that defined the organism's macromolecular individuality (p. 255).

How did Varela view self in relation to mind? Varela (1999) described the mind as phenomenology in action. Viewed from both the first-person and third-person perspectives, he situated behavior in a specific cycle of operation where the locus of the mind emerged through a distributed process within its organizational closure. Mind was an aspect of a pattern in flux in which our concrete biophysical being lives. As embodied selves in dynamic equilibrium, we continually emerge in interactions of constituents and interactions of interactions. Varela and Cohen (1989) viewed the body as the locus where the corporal ego emerges such that the ego gives rise to a sense of self in which this *selfless self* takes on a form so that it looks like our experience inside. Experience continuously shapes this dynamic core at all levels of reciprocal causality through the organizational complementarity of the circular closure of its nervous, hormonal, and mechanical pathways. Varela (1991) conceptualized the organism's identity as an autonomous self, its constitution, nature, and mode of existence as a meshwork of selfless selves:

> Thus we need to deal with a multiplicity of regional selves, all of them having some mode of self-constitution, and in their overall assemblage giving rise to an organism. Accordingly, I want to invoke here the following "regional" selves: (1) a minimal or cellular unity, (2) a bodily self in its immunological foundations, (3) a cognitive perceptuo-motor self associated to animal behavior, (4) a socio-linguistic 'I' of subjectivity, and (5) the collective social multi-individual totality. In all these regions we are dealing with levels and processes where an identity comes about—not as substance, but as movement—and whose fabric of articulation *is* the organism. To efface the multiplicity of this meshwork is a source of confusion. (p. 80)

Varela et al. (1991) argued that cognitive science does not distinguish between the idea or representation of the self and the actual basis of that representation, which is the individual's grasping after an ego-self, nor does it take seriously its own findings of the lack of self, which is rooted in not having a disciplined method for examining human experience. Their concept of embodiment provided for the emergence of a global state among resonating neuronal ensembles in which the autonomic, neuroendocrine, and limbic systems, as autopoietic networks or aggregates of resonant neurons, have patterns of activity that are altered by experience—the phenomenological life world, intentionality, and attention through a process of becoming that is conditioned by its past. Knowledge is thus the result of ongoing subjective interpretations that emerge from our capacities of understanding rooted in the structures of our biological embodiment that are enacted within the domain of consensual action and cultural history.

Prereflectivity and Embodiment

What is the relationship between prereflectivity and embodiment? We are proprioceptively aware of a bodily aspect to human subjectivity from the inside, which makes cognition possible, but perceiving is something that we do, rather than what happens to us (Noë, 2004). As the enactive approach reveals, perception is not only a process in the brain but an activity on the part of the person as a whole. Embodiment thus plays a central role in structuring human experience, cognition, and action. We are both mindful of bodily sensations and we have a sense of ownership built into the prereflective structure of experience that does not require conscious perception or judgment to recognize in awareness or introspection. We live not only along the lifespan, horizontally, but in an ever expanding and contracting experience of states along a vertical plane in the immediate moment (Gordon, in press a, b, 2012, 2013, 2007). We are prereflectively aware as our consciousness expands and contracts with our breath in moment-to-moment subjective experience. However, the experience of self, the quality of *mineness* or *for-me-ness*, self-as-subject, sense of self, or prereflective self-consciousness, pertains to our consciousness of self as the given subject of experience, which remains constantly present while other aspects of phenomenal experience change (Legrand, 2007).

Prereflective self-consciousness is the primary feature of first-person experience or consciousness of the self-as-subject that is not taken as an intentional object. It is characterized by a mode of givenness that *feels like something* for the experiencing subject that is not reducible to phenomenal experience (Legrand, 2007; Zahavi, 2008). I would argue that pure experience is broken down into categories by the senses and experienced as already intrinsically subjective. We do not make sense of conscious experiences that occur to Searle's (2005) "objective' self." There is no screen between our senses and the things from which our senses would otherwise have formed an immediate impression. The distinction between subject and object is not ontological, but phenomenological. However, while I would agree that *self* cannot be an object of experience, *body image* can be an intentional object of consciousness (Legrand, 2007), and *body schema*, which is neither an intentional object of consciousness nor a partial representation of the body, can be viewed as an integrated set of dynamic sensorimotor principles that organize perception and action in a subpersonal and nonconscious manner (Thompson, 2005).

As Legrand (2007) points out, the distinction between body schema and body image leaves out a fundamental form of bodily experience, namely, *prereflective bodily self-consciousness*—or consciousness of the body as subject—the dynamic link between outward perception and inward feeling in ones encountering one's own bodily sentience directly. "Prereflective bodily self-consciousness is evident in touch, for we not only feel the things we touch, we feel ourselves touching them and touched by them" (Thompson, 2005, p. 412). Prereflective consciousness can thus be thought of as the ground to which self-consciousness is anchored (Legrand).

In fundamental agreement with this view of embodiment, Merleau-Ponty (1945/1962) argued for the primacy of perception over cognition, and the unity of experience. He posited that the essence of perception is an objective phenomenon tied to exploratory and goal-directed or intentional movement grounded in the subjective, sense experience of the body. Following from Descartes and Kant, he defined *sensation* as a unit of experience—a phenomenological gestalt or unified field that is "intentional, co-existence or communion" (p. 248), in which the whole system of experience—"world, own body, and empirical self—are subordinated to a universal thinker charged with sustaining the relationship between the three of them" (p. 241). Behavior is thus a projection into the world toward an intentional object or a goal. However, the relationship between self and the world, as Merleau-Ponty correctly points out, is not subject to object, but "the world, which the subject itself projects" (p. 430). Although the unity of experience is established experientially and phenomenologically, there have been many attempts to distill the ingredient(s), which are responsible for this unity. There are many opinions (e.g., De Preester & Tsakiris, 2009; Di Paolo, 2009; Froese & Fuchs, 2012; Langfur, 2012; Melzoff & Moore, 1977, 1983, 1989). According to De Preester (2007), there are sufficient physiological, clinical, and conceptual arguments for a stratification of the body into body image and body schema in which the subjective perspective constitutes *itself* rooted in the *in-depth body*. In agreement with De Preester's (2005), I too assert that the body, as the ground of subjectivity, objectivity, and intersubjectivity, requires a multi-perspectival, interdisciplinary approach to embodiment rather than only observing the object of representation associated with subjective feelings, emotions, and self-awareness in the brain.

What is embodiment? Embodiment refers to the bodily aspects of human subjectivity, the biological and physical presence of our body as a necessary precondition for the experience of emotion, language, thought, and social interaction. It is our kinesthetic awareness of the body as the vehicle through which we experience the sensorimotor, perceptual, and nonconceptual lived world. This is not a cognitive understanding of self in the world, but a proprioceptive, nonconceptual awareness (Fridland, 2011) that is tacit, prereflective, and intersubjective.

This awareness may in fact constitute the beginning of the newborn's primitive body image. Melzoff and Moore (1977, 1983, 1989) found that newborn infants remember and imitate gestures, i.e., open mouth and tongue protrusion (echolalia), once the adult who demonstrates this behavior vanishes from their perceptual field. As Gallagher notes, the study of mirror neurons in the Macaque monkey (Fadiga, Fogassi, Pavesi, & Rizzolatti, 1995; Gallese, 1998, 2001; Gallese, Fadiga, & Fogassi, 1996; Grafton, Arbib, Fadiga, & Rizzolatti, 1996; Rizzolatti, Fadiga, Gallese, & Fogassi, 1996) suggests that the mechanism of imitation is prenoetic because these neurons function in Broca's area from birth. According to Gallagher, the infant's sense of self starts out closer to an embodied sense than to a cognitive or psychological understanding because intermodal translation between the visual and motor systems appear to be operative in the newborn. In agreement with Gallagher, I would argue that the newborn's *proprioceptive self* or embodied perception of self in imitation is intersubjective, "it indicates a rudimentary differentiation between self and non-self… a bare framework of the self that is based on an innate system of embodiment" (2005, pp. 83–84). Likewise, Langfur (2012) presents empirical evidence to suggest that an infant first becomes aware of him or herself as the focal center of a caregiver's attending through neural matching (kinesthesis of the infant's gestures is experienced by the infant as what the caregiver is focusing on). However, this theory does not account for the infant's awareness as agent, which occurs with the acquisition of language.

Recent theories of embodied subjectivity, intersubjectivity, and social cognition claim that individuals exist on the most fundamental level in relation to others given a history of relationship through a joint process of sensemaking (Di Paolo, 2009).

In their theory of the "extended body," Froese and Fuchs (2012) argue that "inter-bodily resonance between individuals" (p. 205) gives rise to self-sustaining interaction patterns that go beyond the behavioral capacities of isolated individuals. Caracciolo (2012) argues that socioculturally mediated meaning-making can be accounted for in enactivist terms through the production and interpretation of narratives. De Preester (2005) argues that body image and schema are embodied in the structuring effect of neuronal organization, bodily experience, and the formation of the subject. I speculate that the intrauterine environment provides the developing fetus with his or her first intersubjective, co-creative, consensual, organismic experience from which maternal hormonal, autonomic, and affective states are imprinted during the period of gestational bonding.

The Embodied Sense of Self

The nature, structure, and reality of the self have been argued about by philosophers for millennia. While "the self" is used as a variable in cognitive research correlated with activity in particular regions of the brain, this only serves to accomplish a transformation of "the self" from an ephemeral theoretical entity to something concrete and measureable (Zahavi & Roepstorff, 2011). In reality, self or the sense of self cannot be an object of itself. Human subjectivity, embodied as felt sense or consciousness of self, is a multidimensional stream of thoughts, perceptions, imaginations, representations, memories, and emotions that are associated with a global state among resonating neuronal ensembles that is continually shaped by the perception of the observer (Gordon, 2013, in press, a, b).

Myers (1892, 2001/1903) proposed the concept of a *subliminal self,* in which consciousness is a spectrum of states ranging from the psychopathic to the transcendent. Myers (1886) described the ability of this subliminal unconscious to express itself as a mythic and poetic imperative. The *mythopoetic* function of the unconscious enabled the conceptualization of reality beyond language through the ineffable symbol in which the unconscious took on a transcendent dimension. James (1890) described consciousness as a stream, a field with a focus and a margin, a plurality of waking and subliminal states (1902), and pure experience embodied in feeling and sensation (1912). Varela's (1999) specious present, borrowing from James (1890) and others linked the "sense of self" with our experience of the stream of consciousness (Gordon, 2009a). Varela, as well as James (1912), argued that the underlying nature of consciousness or the self is a unified field of pure experience with no content other than itself, where all processes of representation are fluctuations or qualified states of this underlying field (Gordon, 2009b). James's concept of pure experience was but another name for feeling or sensation. Its "purity" was relative to the amount of unverbalized sensation that it still embodied (1912, p. 94).

James (1892/1961) saw the mind and body as a fluid, integrated whole prior to the subject-object dichotomy; the self being constituted through their interaction joined at the level of lived experience. He divided the self into an objective known empirical ego—the *me*, further divided into its material, social, and spiritual aspects, and the subjective knower, pure ego, or the *I*. The spiritual me was "composed of the more active feeling states of consciousness; the core and nucleus of our self, a direct revelation of the living substance of the soul" (p. 43), while the I was "the agent, soul, transcendental ego, spirit; or thinker behind the passing state of consciousness lending unity to the passing of thought" (p. 63). The identity found by the "I" in its "me" was only a loosely construed thing, an identity "on the whole" (p. 72) that was divided into mutations of the self based on alterations of memory. Experience had no inner duplicity between subject and object. In his theory of mind-stuff, James posited a relationship between the bare phenomenon or the immediately know thing and the phenomenon of changing brain states or the pulse of consciousness that is cognizant of its object. He argued that the entire brain process *is* the state of consciousness, the soul a medium upon which these processes combine their effects. But, how, or why, no mortal may ever know.

Embodiment of Emotion. James' (1884) theory of emotion is found in his essay "What is an Emotion?," in which he argues for a visceral basis for the emotions as apperceived physiological responses to our perceptions and describes embodiment as the feelings of innervation and the proprioceptive impulses associated with our breath. Emotions were the feelings associated with the immediate perception of an exciting fact as well as the muscular and visceral changes: "*... bodily changes follow directly the* PERCEPTION *of the exciting fact... our feeling of the same changes as they occur IS the emotion*" (pp. 189–190). For James, emotion was purely visceral; it originated in the body and was not cognitive. In 1884, both James and Lange published similar theories:

> An object stimulates one or more receptors, afferent impulses pass to the cortex and the object is perceived; thereupon currents run down to muscles and viscera and alter them in complex ways; afferent impulses from these disturbed organs course back to the cortex and when there perceived transform the 'object-simply-apprehended' to the 'object-emotionally-felt.' (Cannon, 1931, p. 281)

Challenging the James-Lange theory, Cannon's study of the flight or flight response proposed that visceral changes were to too slow to be a source of emotional feelings, but instead the hypothalamus and its neuronal connections to the brainstem, through the secretions of the pituitary gland were the seat of emotion. The thalamocortical pathways were the coordinating center for emotional expression and affective experiences sent to the viscera and skeletal muscles.

Following Cannon's work, Papez (1937) proposed that components of the limbic circuitry form part of a multisynaptic pathway, classically known as Papez's circuit, and are responsible for the expression of emotion. He proposed that the mechanism of operation for emotional regulation began at the hypothalamus (adaptive function/human drives); the cingulate cortex projects to the hippocampus (memory), and the hippocampus projects to the hypothalamus, by way of the fornix. Papez thought that hypothalamic effects reached the cerebral cortex through the anterior thalamic nuclei, which relays sensory impulses such that a stimulus affects the thalamus, the sensory cortex, and the cingulate cortex—creating a feeling— then the cingulate cortex signals the hippocampus and the hypothalamus, leading to a bodily response.

However, Papez was missing an important component of emotional processing—the human amygdala.

Offering evidence that emotion occurs throughout the organism, Pert (1999) proposed that the emotional brain is a somatosensory, non-synaptic, neurochemical gestalt providing an exchange of information that is not confined to the classical locations of the amygdala, hippocampus, and the hypothalamus, but instead, as Schmitt, molecular biologist at MIT discovered, is a parasynaptic or secondary parallel system where chemical substances circulated through the body to reach their target-cell receptor. What we experience as an emotion or feeling was thus a mechanism for activating a particular neuronal circuit simultaneously throughout the brain and body affecting behaviors that involve the whole person. These conclusions were supported by the work in biofeedback on the conscious control of autonomic states by Green, Green, and Walters (1970) who found that change in physiological state is accompanied by change in mental-emotional state, conscious or unconscious, and the converse is also true.

What we understand today, however, is that the neurophysiology of emotion involves bidirectional communication between the cerebral cortex and the limbic system or *corticolimbic system*. The orbital and medial prefrontal cortexes[13] in the anterior frontal lobe communicate with the limbic structures (amygdala, hippocampus, mammillary bodies, and basal ganglia) and the hypothalamus in the processing of memory, the regulation of food-mood responses, attention-alertness, and human drives. These systems communicate by up- and downregulating the neurotransmitters GABA and glutamate, which inhibit or excite arousal, while serotonin, dopamine, and norepinephrine combine their effects to integrate thought, emotion, felt sense, and memory through the symbols of language and mental imagery.

The hypothalamus, master gland and internal watcher of the homeostasis of the organism, located below the thalamus on both sides of the third ventricle, unites the nervous and hormonal systems. Through its connection to the pituitary gland, the hypothalamus regulates the endocrine system, governs the autonomic nervous system, and communicates via a two-way pathway with the limbic system and the cerebral cortex. The limbic system is then informed by sensory input through the fast relay nuclei of the thalamus and the amygdala and the slower pathways of the parahippocampal gyrus.

To briefly return to the missing element in the Papez circuit, the amygdala, which processes aggression, fear, memory, sexual arousal, and pain in moment-to-moment subjective experience, we now know that the amygdala has bidirectional connections with the hypothalamus and receives signals from all parts of the limbic system and the neocortex, especially the auditory and visual association areas. Then the hippocampus, responsible for decision-making, determines the importance of sensory information and transmits signals that allow verbal and symbolic memory to become long term. Given that spatial resolution in neuroimaging presently limits our ability to make claims about localization of function for subdivisions of the human amygdala, neuroscientists, extrapolating from animal studies, have discovered a relationship between the mind and the brain: it is based on the person's perception of and ability to find meaning and significance in their experience.

Perception of the Observer

Before delving into an explanation of the psychoneurointracrine system, I will provide a brief review of empirical research on the amygdala demonstrating a relationship between the mind and the brain. Here I will argue that living systems do not exist in a linear chain of cause-and-effect relations, but within context significance in relation to meaning. The amygdala's sensitivity to emotional meaning imbues experience with affect shaping the content of perception in moment-to-moment subjective experience.

The amygdaloid complex, a group of nuclei and cortical areas with unique cytoarchitecture, contributes to enhancement of memory formation for emotionally arousing events and projects to sensory pathways; then contextual factors (set and setting) determine where feedback is exerted, activating systems of attentional control in subcortical nuclei and the medial prefrontal cortex. According to Phelps (2006), the amygdaloid and hippocampal complexes are linked to two independent memory systems: the amygdala modulates the encoding and storage of hippocampal-dependent memories by forming episodic representations of emotional significance and interpretation of events. The hippocampus influences the amygdala's responsiveness.

Risold and Swanson (1996) showed how the hippocampus and the amygdala give rise to extensive topographically organized projections to the hypothalamus,[14] which modulate the autonomic processing of fear, associative learning, and attention. Holland and Gallagher (1999) found evidence linking the amygdala to the processes of attention, reinforcement, and representation in food-motivated associative learning. For example, when subjects assigned meaning to neural representations of sensory experience in experiments of fear conditioning, the amygdala mediated affective, autonomic, and instinctual responses that integrated subjects' learning about emotionally relevant stimuli and environmental challenges. Emotional context (evaluating meaning) altered behavioral and biological responses when subjects approached or avoided a stimulus with recruitment of the amygdala (Hare, Tottenham, Davidson, Glover, & Casey, 2005). Canli, Zhao, Brewster, Gabrieli, and Cahill (2000) also found a positive correlation between activation of the amygdala during encoding and subsequent memory, supporting the view that activation of the amygdala reflects moment-to-moment subjective experience with the enhancement of memory and emotional intensity.

In association with the prefrontal cortex,[15] the amygdala plays a role in perceptual integration during emotionally significant experiences with either positive or negative valence[16] (Ochsner, Bunge, Gross, & Gabrieli, 2002; Ochsner et al., 2004, 2009). Phelps, Ling, and Carrasco (2006) suggested that emotional cues bias attention to neural locations that enhance the perceptual processing of subsequent stimuli. The effects of emotion on attention and their conjoint effect on contrast sensitivity showed that emotion enhances sensitivity irrespective of attention and potentiates the effect of attention on contrast sensitivity. The amygdala was more responsive to fearful eye whites than happy eye whites when presented in a masking paradigm that mitigated subjects' awareness of their presence and aberrant nature (Whalen et al., 2004). More recently, Vuilleumier (2009) explored how the amygdala integrates subjective experience from internal and external sources through its projections to output systems modulating autonomic, sensory, motor, and cognitive processes, and reciprocal actions between attention and emotion regulate the access of sensory inputs to conscious awareness.[17] These studies collectively show that the amygdala's emotional responsiveness and attention to stimuli is based on the person's perception of experience.

With regard to the study of personality, Canli (2008) argued that traits modulate activation of the amygdaloid complex during the conscious and unconscious processing of emotion. His review of cognitive, neural, and molecular genetics with respect to trait anxiety and neuroticism found that individual differences predicted the response of the basolateral amygdala to unconsciously processed fearful faces. Likewise, Etkin et al. (2004) found that responses to threat-related stimuli were influenced by both conscious and unconscious processes: unconscious processing modulated activity in the basolateral amygdala and conscious processing modulated activity in the dorsal amygdala. While activation of the dorsal amygdala to conscious stimuli appeared consistent across subjects independent of trait anxiety, activity in the basolateral amygdala to unconscious stimuli predicted differences in trait anxiety. These studies suggest a biological basis for unconscious processes characteristic of anxiety. Further, it was found that individual differences in the transporter gene for serotonin modulates affective processing (Canli, Congdon, Constable, & Lesch, 2008) and thus, quite possibly, the perception of the observer. Does this observer have a growth-oriented, spiritual dimension, or neurophenomenological self?

Growth-Oriented Dimension of the Person

Berdyaev (1944, 1951) depicted his spiritual awakening to the present as "nothing that could be described as an orderly development marked by distinct stages of growth, but instead a series of illuminations and crises made intelligible in intuitions" (p. 82). He believed that the person, as a spiritual being, was not outside of noumenal reality, but instead came to know the essence of the divine through the experience of creative ecstasy, intuition, and timelessness in existential time. Existential time is the individual's subjective, qualitative experience in which time is dependent on one's inward change in the intensity of the moment. Existential moments are a dynamic fusion of experience within us, resulting in a dialectical process, out of which the sense of self emerges (May, 1975).

James (1902) posited that transcendent experience was the vehicle for the transformation of personality and the discursive intellect. Beyond the threshold of waking consciousness, he contended, was a connectivity with the unconscious where deep emotions expanded to give an individual a feeling of union with the divine. The subconscious or subliminal region was the doorway to transforming experience, as it opened the entire range of states beyond the margin. James proposed a metaphysics of experience based on changing mental states and the stream of consciousness, a way to explore mystical consciousness as a symbolic language of the subliminal self (Taylor, 1996).

Bühler (1968) posited that healthy personalities were active mediators of their own existence, motivated to maintain homeostasis, change, and growth to fulfill biological and psychoemotional needs and spiritual values. Her work in child and adolescent psychology provided a biologically based theory of personality and self-realization as the goal of human life across the lifespan. Through the integration of its "reality-transcendent orientation" and "subconscious depth" (pp. 18–19), Buhler believed that the self, as a subconscious core system of personality, integrated, directed, and actualized basic human tendencies through creative expansion, existential-will-to-meaning, and intentionality. Development was the "continuous drive to establish and maintain homeostasis [creating] change toward new possibilities… the restitution of deficiencies and the upholding of the complex system's internal order" (p. 17).

In his theory of metamotivation and the biological rooting of the value-life, Maslow (1971) posited that spiritual life (spiritual, religious, philosophical, transcendent, and axiological) is rooted in the biological nature of the species. He viewed spiritual life as having an intrinsic, instinctoid, or hereditary basis in which the highest aspiration of mankind was introjected or unconsciously internalized. In self-actualizing people, this embodied, psychological self became larger than its biological entity through identification with the higher self, the highest values, the nonself, and nature. Maslow described the instinctual motivation to experience ultimate reality or self-transcendence through peak or mystical states once the individual's hierarchical needs for physiological, security, belonging, esteem, cognitive, aesthetic, and self-actualization were met.

Jung (1969) defined consciousness as an event taking place outside in the cosmos that simultaneously produced an internal image as it became conscious. Consciousness did not create itself; "it welled up from unknown depths like a child that was born daily out of the primordial womb of the unconscious" (pp. 569–570). Jung (1971) portrayed the adult as an eternal child who is always becoming and never complete, whose conscious realization or self-actualization through individuation was the aim of human development in the second half of life. Individuation, the attainment of psychic integration, arose from conflict and collaboration between consciousness and the unconscious through the symbolization and mythologization of the transcendent function, which released things from their perceived and conceived associations by recognizing them as delimited representations to our faculties: sensation, thinking, feeling, and intuition. The process of individuation through active exploration of the unconscious uncovered and facilitated the person's potential wholeness.

Jung's typology served to integrate the archetypes of the personal and collective unconscious through complementarity, recognition, and evaluation of sense perceptions or feeling tones, which were susceptible to personification. His ideas were build on the idea that it was possible to experience a higher more refined state of awareness than the normal waking state. If the unconscious could be recognized as a co-determining quantity along with the conscious, the center of gravity of the total personality shifted from the ego as the center of consciousness and instead became located in a hypothetical point between the conscious and unconscious called the self. The path to this experience lay in active exploration of the unconscious. Integration of the ego and the self by means of an integration of the opposites occurred and the self became the center of personality. Jung also believed that the primordial roots of fantasy and symbol formation in myth were psychic energy and libido, which served a compensatory role, providing a bridge between the person's instinctual and spiritual nature. While he believed that instincts were physiologically grounded in the somatic processes of the neural system, the psychoid nature of archetypes extended beyond a neurophysiological basis into the general dynamical pattern of matter and energy.

This growth-oriented transcendent dimension manifests through our personal myths or narratives. Jung's embodiment of nondiscursive symbols or archetypes and Myers's language of the subliminal self are but two examples. Feinstein, Krippner, and Granger (1988) argued that "personal myths are internalized, organizing models that shape perception, understanding, and behavior," and in this sense are "rooted in an individual's biochemistry" (p. 182). Sansonese (1994) described myth as an "esoteric description of a heightened proprioception" (p. 36) rooted in the nervous system. Joseph Campbell depicted the *Hero with a Thousand Faces* (1949). I strongly speculate that this mythopoetic function of the unconscious is built into the human species through psychoneurointracrine mechanisms, and the HPG-HPA axes represent a point where Jung's *imago* and *persona* converge.[18]

Theory of Psychoneurointracrine Autopoiesis

As the reader will recall, psychoneurointracrinology is the study of psychological, neurological, and intracrinological processes forming a mind-brain continuum within the person. Psycho (psychological) refers to the psyche, self, soul, mind, and consciousness. Neuro (neurological) refers to the composition and reactions within the nervous system. Intracrine (intracrinological) refers to the intracellular biosynthesis of steroids, the binding of receptors, and the formation of enzymes that catalyze the creation of hormones within the cell. Intracrinology involves the allosteric enzymatic regulation of steroids that bind to peripheral target sites and modulate feedback from downstream or feed-forward from upstream substrates. A steroid's biological activity depends on its ability to interact with a specific binding site on a corresponding receptor. My theory is that the up- and downregulation of a receptor is autopoietic or self-regulated by the person's perception of experience, sense of self, and moment-by-moment existential-phenomenological state.

To follow my argument concerning how the enactive approach to cognition and the mechanism of stress integration lend support to my theories of embodied cognition, it is important to differentiate a few biochemical facts: first, the HPG (sexual) axis is critical to the development and regulation of reproduction, immunity, and aging; second, the HPA (adrenal) axis is a critical adaptive system that maximizes

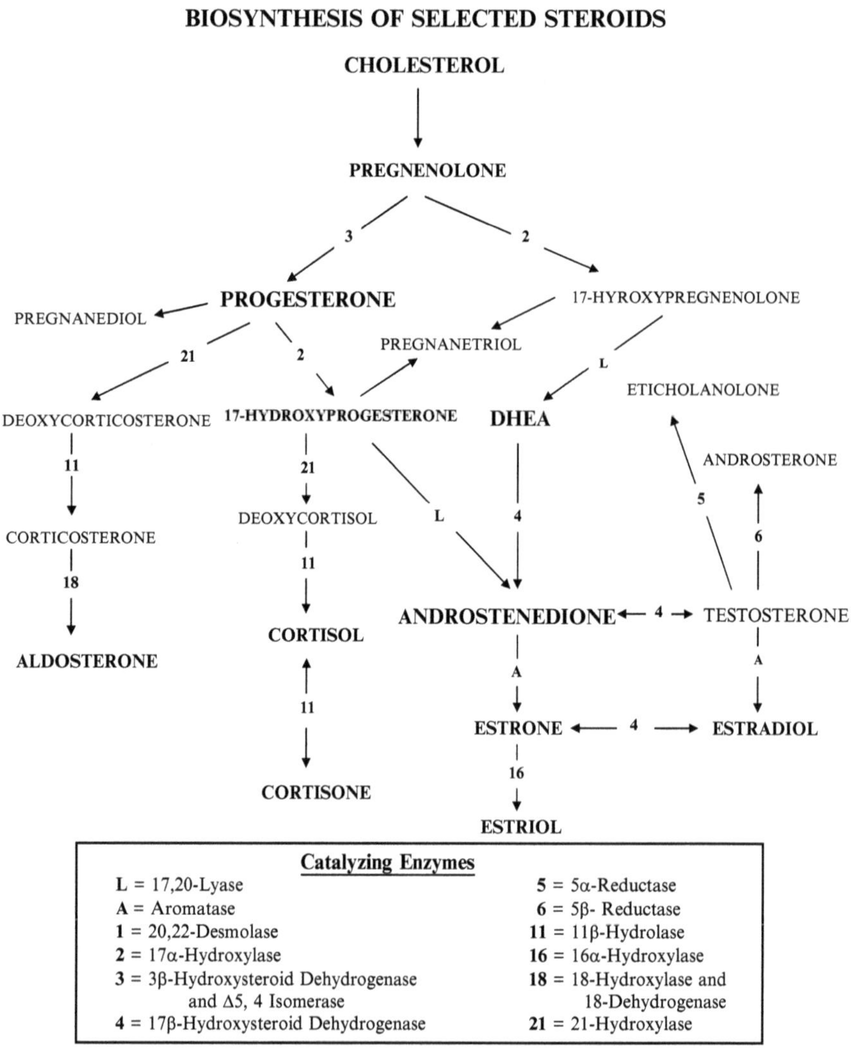

Fig. 1 Biosynthesis of selected steroids. Biosynthesis and metabolism of steroid hormones synthesized from LDL cholesterol and pregnenolone, produced in the ovaries, testes, and adrenal glands. Estrogens are synthesized from DHEA. Glucocorticoids (cortisol and cortisone), and the mineralocorticoid, aldosterone, are synthesized from progesterone. As DHEA produces less estrogen during the follicular phase from its metabolism through androstenedione, progesterone and 17-hydroxyprogesterone may be recruited to metabolize to estrogen via enzymatic cross-conversion utilizing 17α-hydroxylase and 17, 20-Lyase—contributing to deficiencies in progesterone and adrenocortical metabolites (Copyright 2009 by Susan Gordon, Ph.D.)

survival potential in the face of physical or psychological challenge via the fight-or-flight response to stress.[19]

Figure 1, "Biosynthesis of Selected Steroids," represents the sexual and adrenal steroids synthesized from LDL cholesterol and pregnenolone present in the ovary, testes, adrenal gland, and the brain (via the same numbered enzyme systems). On the right side of the chart are sexual steroids synthesized from dehydroepiandrosterone (DHEA): the estrogens (estrone, estradiol, and estriol) and the androgen testosterone. On the left side of the chart are adrenal steroids synthesized from progesterone: the glucocorticoids; cortisol and cortisone, which regulate blood sugar, carbohydrate metabolism, and inflammation; and the mineralocorticoid, aldosterone, which controls fluid balance and blood pressure by regulating electrolytes.

Receptor sensitivity, enzyme titer, and availability of hormonal substrate modulate the biosynthesis of progesterone and DHEA, whose balance plays a major role in the ability to stabilize nerve function and adapt to stress. DHEA, converted from DHEAS (its storage form in peripheral circulation), is both synthesized by the adrenal gland and created de novo by neurosteroids in the brain, which serve primary and secondary roles in adrenocortical function.

DHEA, as precursor to the estrogens, regulates affective arousal via cortical-subcortical control through the HPA circuitry, as well as the synthesis of neurotransmitters, and the receptor sensitivity of neurosteroids. How is this possible? The brain's glial cells synthesize the neurosteroids, pregnenolone, progesterone, DHEA, and their sulfate esters by the same enzyme systems that ovarian and adrenal tissues possess.[20] Neurosteroids alter neuronal function through their influence on transmitter-gated ion channels mediating depolarization and the release of neurotransmitters in the central and peripheral nervous system. They effect cognition (mood, memory, and learning) by exerting biological actions in the brain through either genomic (nuclear steroid receptors) or non-genomic (neurotransmitter receptors) pathways by displaying barbiturate (GABA) or excitatory (glutamate) properties (Jain, Burade, & Subhedar, 1995).

With regard to the mechanism of stress integration, as shown in Fig. 1, when a person experiences either reactive or anticipatory stress, the HPA axis is activated. Reactive stress refers to a real threat to homeostasis that is recognized by somatic, visceral, or circumventricular pathways such as changes in cardiovascular tone, respiratory distress, and visceral or somatic pain. However, HPA activation can also occur in the absence of primary sensory stimuli signaling homeostatic disruption and physiological challenge. This represents an effort by the organism to mount a glucocorticoid response in anticipation of, rather than as a reaction to, homeostatic disruption. Anticipatory responses are generated by conditioning (memory) or by innate species-specific predispositions (i.e., fight or flight, recognition of predators). This is important because the "reactive-anticipatory" distinction is experience dependent. The environment associated with a reactive stressor can be itself conditioned, resulting in an anticipatory response when the conditioned stimuli are next encountered (Herman et al., 2003).

From a biochemical standpoint, when a person who is experiencing stress has sufficient reserves of DHEA, progesterone is rapidly down converted to its adrenocortical metabolites, cortisol, cortisone, and aldosterone to restore homeostasis and the sense of well-being. However, as DHEA becomes progressively deficient due to age and/or chronic stress, adrenal insufficiently may result. This occurs when progesterone is used to metabolize to estrogen via cross conversion of the enzymes 17α-hydrolase and 17, 20-lyase (Fig. 1) resulting in deficiencies of progesterone and progesterone's adrenocortical metabolites, cortisol, cortisone, and aldosterone. If this pattern becomes chronic, it will affect the adequacy of both sexual and adrenal steroids creating a host of stress-induced psychophysical symptoms.

It is important to note that the HPA axis is keenly sensitive and also refractory to fluctuation in gonadal steroids (Herman et al., 2003). The person's ability to modulate homeostasis in the face of stress also influences calcium uptake, which affects central nervous system (CNS) functioning, ability to regulate autonomic sympathetic and parasympathetic arousal, and nutritional status, which affect existential-phenomenological state. How is this possible? The neurointracrine system regulates hormones in peripheral circulation through the modulation of excitatory and inhibitory neurotransmitters that are synthesized by the brain inflecting sympathetic and parasympathetic arousal. More than 50 substances have been postulated or proved to function as synaptic transmitters: small fast-acting molecules such as acetylcholine; amines such as norepinephrine, epinephrine, dopamine, and serotonin; amino acids such as γ-aminobutyric acid (GABA) and glutamate; and nitric oxide, which collectively alter membrane potential modulating the activity of the HPG and HPA axes. Remember that the nervous system does not process information in the computationalist sense, but creates meaning. The subjectively lived body and the living body together have a coordinated response that is experienced and "embodied" throughout the brain and body (Thompson, 2005).

Neurosteroids directly affect cognition by exerting biological actions in the brain through either genomic or non-genomic pathways that govern nuclear receptors for neurotransmitters. They display antagonistic, agonistic, or mixed properties through which they bind receptors for GABA, glutamate, and N-methyl-D-aspartate (NMDA) within the CNS. One of the ways that these transmitters cross the blood–brain barrier is through circumventricular organs (CVOs) such as the median eminence, which modulate hormonal activity at the third ventricle. The CVOs represent an important window through which the brain controls peripheral function that renders CNS neurons responsive to feedback regulation by the secretory products of the target organs they modulate and is fundamental to the way the hypothalamus achieves control over the pituitary and its peripheral targets (Squire et al., 2008).

Of central importance to my theory, the mechanism-regulating CVOs directly affect concentrations of estrogen, progesterone, and DHEA in peripheral circulation affecting the brain's level of neurosteroids. These hormones are known to accumulate at higher levels in the CNS then in the blood or fatty tissue. Estrogen(s) plays an excitatory role in the CNS increasing the sensitivity of the reticular activating system (Goldstein et al., 2005) while decreasing the activity of monoamine oxidase (MAO). Progesterone has an inhibitory role in the CNS, promoting the breakdown of MAO and decreasing its duration of action.

We know that progesterone and its reduced metabolites allopregnanolone and pregnenolone exert a wide spectrum of biological activity on the central and peripheral nervous system such as potentiating GABA transmission by binding to the $GABA_A$ receptor complex at or near barbiturate-binding sites. When the person experiences stress, the adrenal gland converts progesterone to deoxycorticosterone (DOC) and the sedative metabolite, allotetrahydrodeoxy-corticosterone (THDOC)

(Holzbauer, Birmingham, DeNicola, & Oliver, 1985; Majewska, 1987; Majewska, Harrison, Schwartz, Barker, & Paul, 1986). Allopregnanolone, pregnenolone, pregnanediol, and THDOC function as allosteric modulators or agonists of the $GABA_A$ receptor reducing neuronal excitability. Chemically similar to benzodiazepines, these metabolites have anxiolytic, anti-depressant, sedative, and neuroleptic-like effects and share properties with atypical antipsychotic drugs (Holzbauer, 1976; Rapkin, 1999; Rupprecht, et al., 1999; Selye, 1946). My speculation is that consciousness is embodied via these neurohormonal and neuromodulatory mechanisms at the HPG-HPA axes. The theory of psychoneurointracrine autopoiesis is best described as a functional competition for hormonal substrate between sexual and adrenal steroids at all levels of cellular organization.

Theory of Emergent Global States

This theory examines how neuronal projections to the hypothalamic paraventricular nucleus (PVN)[21] from the HPG and HPA axes govern the anticipatory and reactive response to stress[22] and thereby integrate the person's prereflective, autonomic, and subliminal experience. In introducing this topic, it is important to understand a few biochemical facts: first, corticotropin-releasing factor (CRF),[23] a neuropeptide mediator of stress-related endocrine, autonomic, behavioral, and immune responses, is extensively distributed in hypothalamic and extrahypothalamic sites[24] that serves to integrate neuroendocrine activation of the HPA axis by controlling the secretion of adrenocorticotropic hormone (ACTH) from the anterior pituitary. The PVN contains an autonomic division that projects to the brain and the spinal cord, which serves to integrate autonomic and neuroendocrine responses. Widespread distribution of CRF within the CNS integrates neuroendocrine activation of the brain's pituitary-adrenal axis and stress-related autonomic reflexes and behaviors (pain perception, arousal, and motivation) (Strauss & Barbieri, 2009). Second, the "reactive-anticipatory" distinction is experience dependent: the brain generates memory-dependent inhibitory and excitatory traces to control glucocorticoid responses, which are under the exquisite control by limbic brain regions such as the hippocampus, amygdala, and the prefrontal cortex. Hierarchical limbic projections to reactive stress pathways from the medial prefrontal cortex, central amygdaloid nucleus, ventral subiculum, medial amygdaloid nucleus, and lateral septum are limbic system regions known to affect HPA axis activation.[25] While it is believed that none of these regions send direct projections to the PVN, all project to select regions of the brainstem, hypothalamus, and bed nuclei of the stria terminalis (BST), which innervate the medial parvocellular PVN (Herman et al., 2003).

Limbic structures subserving "anticipatory" stress responses modulate PVN activation through interactions with "reactive" stress circuits.[26] The net impact of the stress response is dependent on the ensemble activity of all of these structures. The medial preoptic area sends stress-activated GABAergic projections to the PVN and stands out as a potential interface between gonadal steroids and the HPA axis.

This region, along with subdivisions of the BST, expresses extremely high levels of androgen, estrogen, and progesterone receptors and is thus a major target for gonadal steroid modulation of reproductive function. While it is not known whether steroid-receptive cells project to the PVN, it is clear that the HPA axis is keenly sensitive to gonadal steroids (Herman et al., 2003). Interneurons (relays between neurons) for CRF in the cerebral cortex, the limbic system, and the preoptic area act as excitatory neurotransmitters integrating neuroendocrine activation of the brain's pituitary-adrenal axis with stress-related autonomic reflexes and behaviors (Herman et al., 2003; Strauss & Barbieri, 2009).

More specifically, the ME, a highly vascularized, pituitary stalk, and associated nuclei (raised area on the infundibulum at the third ventricle) connect the hypothalamus to the pituitary and serve as a window for the transport of substances from the blood to the brain forming a crucial link in communication between steroids in peripheral circulation and the CNS with neurotransmitters and neuropeptides. The pathway that controls pituitary ACTH secretion originates in the PVN and terminates at the ME (Strauss & Barbieri, 2009). Axons of GnRH neurons project to the limbic system and circumventricular organs (such as the ME) serving a role as neurotransmitters or modulators in regulating reproductive behavior (Strauss & Barbieri). This CRF-ACTH pathway from the PVN to the arcuate nucleus/ME represents an important link in communication between neuronal axons of the HPG and HPA axis. The intrinsic organization of the PVN is positioned to allow considerable cross talk among cell populations: dendrites of PVN neurons with the cerebrospinal fluid via the third ventricle, release of peptides, and neuronal release and modulation by nitric oxide and carbon monoxide.

In following the biochemistry, you may still be wondering how enactive engagement and the sense of self occur. What I am suggesting is that the mind-brain connection occurs through the body's neurohormonal system. PVN-projecting neurons, localized in region known to receive inputs from somatic nociceptors, visceral afferents, and humoral sensory pathways, integrate corticolimbic pathways. The majority of PVN-projecting neurons are positioned to evoke rapid reflexive activation of the HPA axis and the corticolimbic and gonadal projection systems associated with the CRF-ACTH system. Prereflective, sensory integration and the unity of consciousness come about via the binding of perception: exteroceptors (eyes, ears, nose, tongue, skin) have sense receptors carrying information about the body's external environment to the organism; proprioceptors or interceptors such as baroreceptors in the respiratory tract, digestive tracts, heart, and vascular trunks; chemoreceptors monitoring body chemistry; and mechanoreceptors, nociceptors, and thermoreceptors provide the organism with a sense of the body's internal regulation, kinesthetic position, location, orientation, and movement. States of visceral and proprioceptive knowing or feeling occur in the reticular formation, a neural network connecting the hindbrain and the midbrain, via the hypothalamus, with the cerebral cortex. The hypothalamus, linked via the thalamus to sensory organs, the cortex, the viscera, and the endocrine system, particularly the pituitary, regulates autonomic behaviors (heart rate, blood pressure, temperature, and the release of hormones) and thus the individual's existential-phenomenological state. The person projects, embodies, and enacts this homeostatic global state among neuronal ensembles by "being in the world" and coupling with the environment. This constitutes the growth-oriented dimension of the person and the neurophenomenological self.

Evidence from the Human Menstrual Cycle

The dynamic complexity of the psychoneurointracrine system is evident in the autopoietic regulation of the menstrual cycle. The hormonal cycle, which appears to be governed by feedback mechanisms, is crucially dependent on hypothalamic secretions of gonadotropin-releasing hormone (GnRH) by the pituitary that occur in oscillations of high-frequency signaling. While gonadal steroids vary systematically across the menstrual cycle, they influence neuronal activity and affective processing. As we have said previously, DHEA, the precursor of estrogens in women, intracrinologically governs the catalyzing of enzymes, the synthesis of neurotransmitters, and the receptor sensitivity of neurosteroids. Research on the intracrine metabolism of DHEA supports this assertion (Kuhn-Velten, 2000; Labrie, 1991, 2003, 2004; Labrie, Belanger, Cusan, & Candas, 1997).

Of particular interest, in vivo imaging studies at Brigham and Women's Hospital found that estrogens likely regulate affective arousal in women via cortical-subcortical control through hypothalamic circuitry (Goldstein et al., 2005). Significant differences in activation of the HPA axis in adult women showed attenuation during ovulation and increased activation during the early follicular phase associated with the magnitude of blood oxygenation level-dependent signal changes. During the early follicular phase compared with mid-cycle, the central amygdala, paraventricular and ventromedial hypothalamic nuclei, hippocampus, orbitofrontal cortex, anterior cingulate gyrus, and the peripeduncular nucleus of the brainstem were involved in the stress response. The hypothalamic-pituitary-ovarian (HPO) and HPA axes have been implicated in the neurodevelopmental etiology of schizophrenia (Goldstein, 2006), while menstrual phase modulated affective arousal and prefrontal asymmetry (Hwang, Wu, Chen, Yeh, & Hsieh, 2009).

Research on premenstrual syndrome sheds light on the relationship between affect and arousal. The onset of fluctuant moods in women frequently coincides with rapid change in ovarian steroids and the regulation of neurosteroids and neurotransmitters by MAO (Rapkin, 1999). Noradrenaline and cortisol were shown to vary across the menstrual cycle with the highest values obtained during the luteal phase. Mood and somatic symptoms also show phase-related change: more negative moods in the luteal and menstrual phases and more positive moods in the follicular and ovulatory phases (Collins, Eneroth, & Landren, 1985). Likewise, a clear temporal relationship between mood, physical state, and hormonal phase was demonstrated in self-ratings of well-being reaching their maximum in the late follicular phase and declining throughout the luteal half of the cycle (Sanders, Warner, Backstrom, & Bancroft, 1983). Depressed women, when compared with schizophrenic women, show an elevation in frequency of admission to psychiatric facilities on the last premenstrual day and the first day of the menstrual flow (Abramowitz, Bakerk, & Fleisher, 1982) related to a sudden drop in estrogen and progesterone.

The phenomenon of mood swings is often shown to be related to depleted estrogen, the activity of type A MAO, and the breakdown of serotonin, norepinephrine, and dopamine 48 hours before bleeding; as the level of estrogen drops, MAO remains high; as catecholamines break down, estrogen in the brain is depleted (Luine & McEwen, 1997). There is also a significant positive correlation between estrogen(s) and the plasma level of the serotonin precursor 5-hydroxy-tryptophan (Thompson, Maddock, Ayaland, & Oswald, 1977). Platelet levels of serotonin, estrone, and estradiol in nondepressed women found that mood-enhancing effects of estrogen correlated with increased number of binding sites for imipramine (Sherwin & Surayani-Cadotte, 1990). Interestingly, cases of menstrual and puerperal psychosis implicate estrogen excess and progesterone deficiency with high levels of MAO in the pathogenesis (Brockington & Meakin, 1994). These studies collectively suggest that the net effect of gonadal steroids is related to the interneuronal excitation and global state within the CNS. This research demonstrates how the theory of psychoneurointracrine autopoiesis and emergent global states apply to women across the lifespan, particularly during transitions such as puberty, pregnancy, postpartum, and menopause. During these times, a woman is most vulnerable to the effects of stress resulting in deficiencies of estrogen, progesterone, and DHEA, which contribute to a host of psychophysical symptoms clinically labeled as PMS, PMDD, and menopausal syndrome, when they are often more indicative of the person's dialectical path toward psychospiritual growth, wholeness and self-actualization.

Conclusion

My aim in this chapter has been to introduce the psychoneurointracrine model of the embodied self and explain my theories of enactive cognition and the sense of self using the works of Francisco Varela and contemporary neurophenomenologists. I argued that self has neural correlates in the sexual and adrenal axes of the body, which are responsible for enactive engagement and the development of meaning through their connections to the higher-order functions of the brain. The physiology and function of the brain, energetic components of disease, correlation of symptoms of disease with the emotions, and the effect of beliefs held in the unconscious mind, are important considerations in the clinical assessment of adrenocortical health and psychological well-being. These are predictors of psychoneuroses as well as existential challenge. Limbic and HPA axis dysregulation are features of affective disorders, but also the dialectical response compatible with self-realization. States of affective disease and self-transcendent experiences activate the same self channels; anticipatory HPA pathways operate in both innate and learned response predispositions.

It was hypothesized that adrenocortical progesterone and DHEA levels fluctuate in accord with the sense of self as the boundaries of the personality are integrated.

It is not what happens to a person, but how he or she interprets it that determines its meaning, its outcome, and its myth. The theory of psychoneurointracrine autopoiesis examines how the regulation of a steroid's receptor is modulated by the person's perception of experience. The theory of emergent global states explains how corticolimbic projections of the HPG-HPA axes integrate prereflective, autonomic, subliminal experience in the development of meaning and emergence of self.

The psychoneurointracrine system reveals the observer's pattern of activity that is altered by experience—the phenomenological life world, intentionality, and attention through a process of becoming that is conditioned by its past. Steroids are catalysts of infinitesimal concentrations supporting and sustaining the evolution of human cells. It is their autopoietic capacity to transduce and transform that is the locus of an individual's energic balance. This system reveals the existential-phenomenological state of the organism. It gives primacy to experience as a lived, embodied, intersubjective point of reference for self in the world. It addresses the ontological priority of consciousness and the perception of the observer. It explores emotion as the ground from which cognition occurs and prereflective awareness (preconceptual or pre-languaged bodily knowing), which exists prior to felt sense and allows for meaningful experience on the level of conceptual awareness. This model of the neurophenomenological self extends knowledge of the mind-brain relationship and the growth-oriented dimension of the person.

Notes

1. The growth-oriented dimension refers to an expanded definition of consciousness and personality that focuses on self-reflective, existential-phenomenological, spiritual-transcendent, and subliminal aspects of the person. The focus on the "self" is not on its measurement, but on its direct experience through the actualization of its potential and its striving toward health as intrinsic to human motivation. For example, James (1902) explored the subconscious as the doorway to transforming, mystical experience (see Taylor, 1984, 1996, 2009). Jung (1971) explored individuation of the self though his analytic and depth psychology. Maslow (1954, 1971) addressed stages of self-actualization and self-transcendence in his hierarchy of human needs. Bühler (1968) referred to a subconscious core system of personality. Watts (1975) taught Buddhist psychology to the West. For an historical account, see also Bugental (1963); Ellenberger (1970); Frankl (1966); May (1975); May, Angel, and Ellenberger (1958), and Rogers (1964).
2. The term *enactivism,* first introduced by Varela et al. (1991), defined a paradigm in cognitive science that can be traced back to the program in the biology of cognition proposed by Maturana (1975, 1988a,1988b), further developed by Maturana and Varela (1987; Varela, 1979a, 1981). The enactive approach unified the ideas that (1) living beings are autonomous agents that actively generate and maintain their identities, thereby *enacting* or bring forth their own cognitive domains; (2) the nervous system, as an autonomous system, actively generates and maintains its own coherent and meaningful patterns of activity according to its operation as an organizationally closed or circular and reentrant sensorimotor network of interacting neurons. The nervous system does not process information in the computationalist sense, but creates meaning; (3) cognition is a form of embodied action (cognitive structures emerge from recurrent sensorimotor patterns of perception and action); and (4) a cognitive being's world is not a prespecified, external realm, represented internally by its brain, but a

relational domain enacted by that being's autonomous agency and mode of coupling with the environment (Thompson, 2005).

3. To understand the complex interplay between brain, body, and world, the neuroscience of consciousness used the tools and methods of nonlinear dynamic theory. Traditional notions of representation and computation that presume a one-way causal explanatory relationship were deemed inadequate; therefore a two-way or reciprocal relationship between neural events and conscious activity was needed (Thompson & Varela, 2001, p. 418).
4. Complex systems are heterogeneous entities with multiple levels of collective structure and organization having unique features that are often overlooked by science including (a) the emergence of superstructures, which dominate the system's behavior and cannot be traced back to the properties of the constituent entities; (b) higher emergent features arising from lower-level interactions; (c) global patterns, which affect lower levels through feedback loops that allow the system to adapt and maintain stability; and (d) robustness against large-scale and potentially disruptive perturbations. The study of complex systems requires observation and analysis at many different spatial and temporal scales to derive theories regarding their behavior and control using an interdisciplinary approach intended to raise universal questions that find expression in widely different systems across a spectrum of disciplines such as biochemistry, neuroscience, and psychology (L'Institut des Systèmes Complexes, 2009).
5. For example, see Barden (2004), Bradley and Dinan (2010), Cubała and Landowski (2006), Herman et al. (2003), Ryan and Herman (2008), and Tsigos and Chrousos (2002).
6. The term "intracrine" refers to peptide/protein hormones that have intracellular actions and/or act within the cell that synthesizes it. Intracrinology is the study of hormones that are synthesized and exert their action in the same cell without exit and reentry (in contrast to autocrine signaling where an effector molecule leaves the cell of its origin before it rebinds to the surface of that cell). According to the intracrine theorem, the link between unsuitable ligands and receptors is an intracellular enzyme that converts an inactive prohormone into a ligand that is able to activate that receptor and to generate a biological response. Intracrine modulation of cellular function relies on intracellular processing of an endogenously synthesized compound, followed by a signal transduction pathway of the finally processed molecule within one and the same cell (Kühn-Velten, 2000).

   For an excellent review of this topic, see Re (2002, 2003), Re and Bryan (1984), and Re and Cook (2006, 2008). With regard to the scientific and clinical implications of the intracrine metabolism of dehydroepiandrosterone (DHEA) into its bioactive metabolites, see Burger (2002), Labrie (1991, 2003, 2004), Labrie, Belanger, Cusan, and Candas (1997), Labrie et al. (1998), Labrie, Belanger, Simard, Luu-The, and Labrie (1995), Labrie et al. (2005), Labrie, Luu-The, Labrie, and Simard (2001), and Simpson (2002).
7. Hormone receptors modulate feedback from downstream or feed-forward from upstream substrates. As highly specific, enzyme-substrate systems, hormones (fat-soluble ligands) exert their actions through nuclear receptor-mediated transcription in which cell surface proteins bind hormones triggering receptor interactions that affect the behavior and plasticity of genes (Burtis, Ashwood, & Bruns, 2011).

   Receptor sensitivity is largely based on the availability of substrate. However, synaptic binding is not a linear process. Ionotropic, metabotropic, and cumulative effects create hormonal interactions, which may not be reversible and have lasting impact on one's ability to homeostatically modulate (e.g., long-term SSRI use decreases receptor sensitivity and population and often leads to addiction).
8. The emphasis in existentialism is that people are free and also responsible for what they make of themselves and their lives. The emphasis in phenomenology is on knowledge and meaning. Existential phenomenology describes how phenomena present themselves in lived experience or the individual's subjectivity in human existence. The European existential-phenomenological psychotherapeutic traditions united by Rollo May and Henri Ellenberger in the 1950s and 1960s span the psychologies of Kierkegaard, Buber, Tillich, Jaspers, Nietzsche, and Heidegger and the humanistic psychologies of James, Allport, Maslow,

Rogers, May, and Bühler, among others. The fusion of these traditions has resulted in a humanistic philosophy whose breadth spans the interdisciplinary fields of language, art, psychology, ethics, epistemology, psychiatry, psychoanalysis, sociology, religion, law, and anthropology. Existential phenomenology is primarily concerned with how we become conscious of meaning through feeling, perceiving, remembering, imagining, willing, and reflecting. The most decisive trait of human consciousness is that it is an embodied consciousness or embodied subjectivity with spatiality, temporality, self-expression, and sexuality. Existential moments are a fusion of experience within us, resulting from dialectical process, out of which the sense of self emerges (May, 1975).

9. The word *cybernetics*, used as early as Plato, described the art of governing. Weiner (1948) introduced the concept as a doctrine of general regulations (i.e., homeostasis), which occur in machines and organisms. Elements of cybernetics (subjective self-regulation: information theory, assimilation, coding, and programming) gave rise to formal explanations in genetics, physiology, neurophysiology, and behavior and a bridge between the sciences (Buytendijk, 1974).
10. Varela and Goguen (1977) defined the phenomenology, circular organization, and operations of internally coherent self-referential systems in which the wholeness of the system is embodied in its organizational closure having interactions in a static circular framework with dynamic stability in relation to environmental perturbations. The wholeness of this system was not defined as the sum of its parts, but as a universal mechanism for its stabilization. Maturana and Varela (1987) clarified that autopoietic systems regenerate and transform themselves by their own mutual interactions through *structural coupling*, a mechanism whereby the nervous system couples sensory and motor surfaces through a network of neurons amidst a changing internal environment.
11. James' references to classical Eastern psychology, specifically to Vedanta, Hinduism, and Buddhism, came through American literary sources such as Ralph Waldo Emerson, the transcendentalist circle, and his father, Henry James, Sr. (Taylor, 1996).
12. Asian contemplative traditions find no world of objects independent of the observer and no actual "self" in consciousness—what remains preconstructively is a world of pure experience. For example, in Indo-Tibetan Buddhism, the individual does not exist independently of the body and mind, but rather in reliance of myriad physical and mental processes that are constantly changing and intersubjective. Likewise, in the *Abhidharma* tradition, where the temporal relationship between consciousness and its object has also been the subject of dispute, it is believed that every moment of consciousness is bound to its object by five mental aggregates, serving as contact between the mind and its object, via a specific feeling tone or an intention toward that object (Varela et al., 1991).
13. The medial prefrontal cortex (mPFC) is involved in the processing of convergent cognitive, emotionally relevant information, and modulating attentional states. It is an important neural substrate for integrating cognitive-affective information and regulating the HPA axis response to emotional stress. Dorsal and ventral aspects of the mPFC differentially regulate neuroendocrine and autonomic PVN outputs (Radley, Arias, & Sawchenko, 2006).
14. This projection system (1) regulates neuroendocrine and autonomic function and ingestive behavior (periventricular zone), (2) modulates reproductive and agonistic behaviors (rostral medial zone), and (3) provides polysynaptic feedback/projections to the hippocampus via the mammillary complex. Limbic projections arising from the amygdala influence hypothalamic function via two pathways: (1) the stria terminalis, which arises from neurons in the amygdala similar to the course of the fornix, and (2) amygdaloid neurons that innervate the hypothalamus via the ventral amygdalofugal bundle, which passes directly over the optic tract to the hypothalamus.
15. The prefrontal cortex constructs reappraisal strategies that modulate emotion processing. Self-focused regulation recruits the medial prefrontal regions, whereas situation-focused regulation recruits the lateral prefrontal regions.

16. Bentley, Vuilleumier, Thiel, Driver, and Dolan (2003) suggested that acetylcholine modulates selective attention and emotional processes through independent, region-specific effects; cholinergic input to the frontoparietal cortex influences the allocation of attention to emotional information.
17. Vuilleumier (2009) offered the following mechanism of action between emotion and attention: (1) the amygdala induces feedback from the lateral and basal nuclei, (2) neural representation of emotional stimuli promote selection by attention, (3) privileged access to competing stimuli, and (4) attenuation by emotional states via gating in the basal nucleus mediated by the orbitofrontal cortex.
18. Jung's (1911/2001) *imago* represents the idealized mental image the person has of himself or herself, how the subject apprehends the self based on intersubjective relations that are unconsciously projected. Jung replaced the term *imago* with the term *complex* and later with the term *archetype* in order to express the idea that it involves impersonal, intrapsychical collective motifs that are situated between the unconscious and consciousness. *Persona* (Greek: mask), one of the five Jungian archetypes (self, shadow, anima, animus, persona), is the face or image (mannerisms or traits embedded deep within the psyche) that are knowingly portrayed by the individual to the outside world. *Persona* evolves from childhood as the individual navigates the stereotypes and norms of society and develops a psychological framework for relating to others.
19. Activation of the HPA axis can occur reflectively in response to physical challenge. Reactive responses are driven by ascending brain systems or circumventricular organs, which send direct projections to the PVN. Anticipatory responses are initiated through comparison of environmental stimuli (instinctual fear of predators) or memories (prior experience with painful stimulus (psychogenic stressors)) (Jankord & Herman, 2008).
20. These neurosteroids have been found in all regions of the rodent's brain in concentrations severalfold higher than in plasma (Baulieu, 1996; Baulieu & Robel, 1998; Robel & Baulieu, 1995).
21. PVN-projecting neurons are localized in region known to receive inputs from somatic nociceptors, visceral afferents, and humoral sensory pathways. The majority of PVN-projecting neurons are positioned to evoke rapid reflexive activation of the HPA axis. Intrinsic PVN-regulatory pathways: (1) the PVN contains a dense capillary plexus affording ready access to steroid hormones; (2) PVN capillaries also synthesize prostaglandin (E2), which can interact with receptors on parvocellular PVN neurons to provide a mechanism for local signaling; (3) though dendrites do not extend outside the nucleus, they ramify extensively within the nucleus affording interaction with other cell populations; (4) neurons within the parvocellular and or magnocellular subdivisions contain cellular machinery to synthesize the gaseous neurotransmitter nitric oxide (NO) and carbon monoxide (CO), both of which affect HPA axis activation; (5) evidence for local lateral innervation of parvocellular neurons by intra-PVN axons suggests an opportunity for synaptic interactions within and across PVN cell groups; and (6) terminals in the medial eminence have access to blood-borne and glial constituents that can be transported to affect growth and plasticity (Herman et al., 2003, p.162).
22. The HPA axis operates in two domains: relatively unstressed conditions in which glucocorticoids secretion undergoes a daily rhythm with peak secretion occurring at initiation of the waking cycle. Activity is controlled by suprachiasmatic nucleus and sets the tone for information processing in the brain. During stressed conditions, there is a real or predicted threat to homeostasis during which the brain initiates an ACTH surge to promote adrenocortical activation. Real or "reactive" stress represents a genuine homeostatic challenge that is recognized by somatic, visceral, and circumventricular sensory pathways with change in cardiovascular tone, respiratory distress, visceral or somatic pain, and blood-borne cytokine factors signaling infection or inflammation. In the absence of a primary sensory stimuli signaling homeostatic disruption and the absence of physiological challenge, there is an effort by the organism to mount a glucocorticoid response in anticipation of, rather than as a reaction to, homeostatic disruption. "Anticipatory" responses are generated by conditioned memory or by innate species-species predispositions (i.e., recognition of predator) (Herman et al., 2003).

23. Regions that directly innervate these neurons are primed to relay sensory information including visceral afferents, nociceptors, and circumventricular organs—promoting "reactive" corticosteroid responses to emergent homeostatic challenges. Indirect inputs from the limbic-associated structures are capable of activating these same cells in the absences of physiological challenges; such "anticipatory" signals regulate glucocorticoid release under conditions in which physical challenges may be predicted either by innate programs or conditioned stimuli. Anticipatory circuits are integrated with neural pathways subserving "reactive" responses at multiple levels. The resultant hierarchical organization compares information from multiple limbic sources with internally generated and peripherally sensed information, thereby tuning the relative activity of the adrenal cortex.
24. The PVN receives its primary input from the (1) nucleus of the solitary tract—norepinephrine and epinephrine relay visceral and somatic sensory information; (2) locus ceruleus—the medial prefrontal cortex and the amygdaloid nucleus relay information related to emotion; (3) raphé nuclei (serotonergic); (4) subfornical organs (status of fluids and electrolytes); (5) hypothalamus (GABA [inhibitory] and glutamate [excitatory] neurons); (6) bed nucleus of the stria terminalis (receptors for gonadal steroids and CRH neurons); (7) thalamus (auditory-sensory input); and (8) glucocorticoids (blood sugar). The medial preoptic area sends stress-activated GABAergic projections to the PVN and is the potential interface between gonadal steroids and the HPA axis.

    The PVN receives direct innervation from extrahypothalamic pathways regulating homeostatic functions: (1) subfornical organs (SFO)-median preoptic nucleus (mnPOA)-organum vasculosum of the laminae terminalis (OVLT) [regulating fluid and electrolyte balance]; (2) norepinephrine (NE), epinephrine (E), glucagon-like peptide 1 (GLP-1) and other neuropeptidergic neurons in the nucleus of the solitary tract (NTS), and parabrachial nucleus (PBN) [subserving relay of autonomic and immune system afferents]; and (3) hypothalamic nuclei [subserving autonomic/metabolic/immune/arousal signals], including the dorsomedial hypothalamus (DHM), medial preoptic area (mPOA), lateral hypothalamus area (LHA), arcuate nucleus (ARC), peri-PVN zone, anterior hypothalamic nucleus (AHN), and the ventral premammillary nucleus (PMV), among others. Projections are GABAergic, glutamatergic, or mixed populations of cells (Herman et al., 2003, p. 155).
25. Direct projections innervated by the limbic system include bed nucleus of the stria terminalis (BST), dorsomedial hypothalamus (DMH), preoptic area (POA), and nucleus of the solitary tract (NTS). Glutamatergic innervation of direct projections occurs: ventral subiculum (vSUB) and medial prefrontal cortex (mPFC); GABAergic innervation occurs in the medial amygdala (MeA) and central amygdala (CeA) (Jankord & Herman, 2008).
26. Limbic sites have minimal direct projections to HPA effector neurons of the PVN. It is believed that hippocampal, cortical, and amygdalar efferents relay with neurons in the BST, hypothalamus, and brainstem to access corticotropin releasing hormone neurons. Hippocampal, cortical, and amygdalar projection pathways show extensive overlap in regions such as the BST, hypothalamus, and brainstem, implying that limbic information may be integrated at subcortical relay sites prior to accessing the PVN.

    Limbic sites also show divergent projections, with the various structures having distinct subcortical targets. All regions express both glucocorticoid and mineralocorticoid receptors, allowing for the modulation of limbic signaling patterns. Glucocorticoid secretion is driven by internally perceived disruptions of homeostasis, which are cued by cardiovascular, respiratory, or visceral stimuli. These disruptions appear to be relayed to the PVN by way of brainstem neurons, located in the region of the nucleus of the solitary tract and ventrolateral medulla. Glucocorticoid responses can be initiated by direct activation of the PVN via nociceptive pathways (pain), recruitment of defenses, or association cued by multimodal sensory stimuli (i.e., fear) (Choi et al., 2007; Herman, Ostrander, Mueller, & Figueiredo, 2005; Jankord & Herman, 2008).

References

Abramowitz, E. S., Bakerk, A. H., & Fleisher, S. F. (1982). Onset of depressive psychiatric crisis and the menstrual cycle. *American Journal of Psychiatry, 139*, 475–478.

Adams, F. (2010). Embodied cognition. *Phenomenology and Cognitive Science, 9*, 619–628.

Barden, N. (2004). Implication of the hypothalamic–pituitary–adrenal axis in the physiopathology of depression. *Journal of Psychiatry and Neuroscience, 29*(3), 185–193.

Baulieu, E. E. (1996). Dehydroepiandrosterone (DHEA): Fountain of youth? *Journal of Clinical Endocrinology Metabolism, 81*, 3147–3151.

Baulieu, E. E., & Robel, P. (1998). Dehydroepiandrosterone (DHEA) and dehydroepiandrosterone sulfate (DHEAS) as neuroactive neurosteroids. *Proceedings of the National Academy of Science, 95*, 4089–4091.

Bentley, F., Vuilleumier, P., Thiel, C. M., Driver, J., & Dolan, R. J. (2003). Cholinergic enhancement modulates neural correlates of selective attention and emotional processing. *NeuroImage, 20*(1), 58–70.

Berdyaev, N. (1944). *Slavery and freedom*. New York, NY: Scribner's.

Berdyaev, N. (1951) *Dream and reality: An essay in autobiography*. New York, NY: Palgrave Macmillan.

Bradley, A. J., & Dinan, T. G. (2010). A systematic review of hypothalamic–pituitary–adrenal axis function in schizophrenia: implications for mortality. *Journal of psychopharmacology Oxford England, 24*(11), 91–118.

Brockington, I. F., & Meakin, C. J. (1994). Clinical clues to the aetiology of puerperal psychosis. *Progressive Neuropsychopharmacology, Biology, and Psychiatry, 18*, 417–429.

Bugental, J. F. T. (1963). Humanistic psychology: A new breakthrough. *American Psychologist, 18*, 563–567.

Bühler, C. M. (1968). The general structure of the human life cycle. In C. Bühler & F. Massarik (Eds.), *The course of human life: A study of goals in the humanistic perspective* (pp. 12–26). New York, NY: Springer.

Burger, H. G. (2002). Androgen production in women. *Fertility and Sterility, 77*(4), S3-S5

Burtis, C. A., Ashwook, E. R., & Bruns, D. E. (2011). *Tietz textbook of clinical chemistry and molecular diagnostics* (5th ed.). Philadelphia, PA: W. B. Saunders.

Buytendijk, F. J. J. (1974). *Prolegomena to an anthropological physiology*. Pittsburgh, PA: Duquesne University Press.

Cai, W. (2012). Between the sense of self and the reality of self. *Phenomenology and the Cognitive Sciences, 11*, 113–118.

Campbell, J. (1949). *The hero with a thousand faces*. New York, NY: Bollingen Foundation.

Canli, T., Zhao, Z., Brewster, J., Gabrieli, J. D., & Cahill, L. (2000). Event related activation in the human amygdala associates with later memory for individual emotional experience. *Journal of Neuroscience, 20*, RC99.

Canli, T. (2008). Towards a neurogenetic theory of neuroticism. *Annals of the New York Academy of Science, 1129*(1), 153–174.

Canli, T., Congdon, E., Constable, R. T., & Lesch, K. P. (2008). Additive effects of serotonin transporter and tryptophan hydrolase-2 gene variation in neural correlates of affective processing. *Biological Psychology, 79*, 118–125.

Cannon, W. B. (1931). Again the James-Lange and the thalamic theories of emotion. *The Psychological Review, 38*(4), 281–295.

Caracciolo, M. (2012). Narrative, meaning, interpretation: An enactivist approach. *Phenomenology and the Cognitive Sciences, 11*, 367–384.

Choi, D. D., Furay, A. R., Evanson, N. K., Ostrander, M. M., Ulrich, Y. M., & Herman, J. P. (2007). Bed nucleus of the stria terminalis subregions differentially regulate hypothalamic-pituitary-adrenal axis activity: Implications for the integration of limbic inputs. *The Journal of Neuroscience, 27*(8), 2025–2034.

Collins, A., Eneroth, P., & Landgren, B. (1985). Psychoneuroendocrine stress responses and mood as related to the menstrual cycle. *Psychosomatic Medicine, 47*(6), 512–527.

Cubała, W. J., & Landowski, J. (2006). Serotoninergic system and limbic-hypothalamic-pituitary-adrenal axis (LHPA axis) in depression. *Psychiatria Polska, 40*(3), 415–430.

De Preester, H. (2005). *Body image and body schema: Interdisciplinary perspectives on the body.* Philadelphia, PA: John Benjamins Publishing Company.

De Preester, H. (2007). The deep bodily origins of the subjective perspective: Models and their problems. *Consciousness and Cognition, 16*, 604–618.

Di Paolo, E. (2009). Editorial: The social and enactive mind. *Phenomenology and the Cognitive Sciences, 8,* 409–415.

Ellenberger, H. (1970). *The discovery of the unconscious.* New York, NY: Basic Books.

Etkin, A., Klemenhagen, K. C., Dudman, J. T., Rogan, M. T., Hen, R., & Kandel, E. R.,…Hirsch, J. (2004). Individual differences in trait anxiety predict the response of the basolateral amygdala to unconsciously processed fearful faces. *Neuron 44*(6), 1043–1055.

Fadiga, L., Fogassi, L., Pavesi, G., & Rizzolatti, G. (1995). Motor facilitation during action observation: A magnetic stimulation study. *Journal of Neurophysiology, 73*, 2608–2611.

Feinstein, D., Krippner, S., & Granger, D. (1988). Mythmaking and human development. *Journal of Humanistic Psychology, 28*, 23–50.

Frankl, V. (1966). Self-transcendence as a human phenomenon. *Journal of Humanistic Psychology, 6,* 97–106.

Fridland, E. (2011). The case for proprioception. *Phenomenology and the Cognitive Sciences, 10*, 521–540.

Froese, T., & Fuchs, T. (2012). The extended body: A case study in the neurophenomenology of social interaction. *Phenomenology and the Cognitive Sciences, 11*, 205–235.

Gallagher, S. (2005). *How the body shapes the mind.* Oxford, NY: Oxford University Press.

Gallese, V. (1998). Mirror neurons: From grasping to language. Paper presented at Tucson III Conference: Towards a Science of Consciousness.

Gallese, V. (2001). The "shared manifold hypothesis:" From mirror neurons to empathy. *Journal of Consciousness Studies, 8*(5–7), 33–50.

Gallese, V., Fadiga, L., & Rizzolatti, G. (1996). Action recognition in premotor cortex. *Brain, 119*, 593–609.

Goldstein, J. M. (2006). Sex, hormones and affective arousal circuitry dysfunction in schizophrenia. *Hormones and Behavior, 50*, 612–622.

Goldstein, J. M., Jerram, M., Poldrack, R., Ahern, T., Kennedy, D. N., & Seidman, L. J.,…Markis, N. (2005). Hormonal cycle modulates arousal circuitry in women using functional magnetic resonance imaging. *The Journal of Neuroscience, 25*(40), 9309–9316.

Gordon, S. (2001). *A theoretical investigation of the relationship between transpersonal self-concept and estrogen dominance.* Unpublished master's thesis. Saybrook Graduate School and Research Center, San Francisco, CA.

Gordon, S. (2007). Psycho-neuro-intracrinology: The embodiment of myth, intentionality, and the spiritual sense of well-being in women at midlife. *Dissertation Abstracts International, B 68/09.* (UMI No. 3285857)

Gordon, S. (2009a, September). The mind and the brain [Review of the book *The Human Amygdala*]. *PsycCritiques – Contemporary Psychology: APA Review of Books, 54* (No.39), Article 9. PsycCRITIQUES database.

Gordon, S. (2009b). Radical empiricism and the neurophenomenology of time. Unpublished working paper.

Gordon, S. (2012, February). Existential time and the meaning of human development. *The Humanistic Psychologist, 40*(1), 79–86.

Gordon, S. (April, 2013). Internal time is autopoietic. [Review of the book *Internal Time: Chronotypes, Social Jet Lag, and Why You're So Tired]. PsycCritiques – Contemporary Psychology: APA Review of Book*, 58 (No.15), Article 4. PsycCRITIQUES database.

Gordon, S. (in press a). A neurophenomenological psychology of self. In R. Brook, C. T. Fischer, & L. Laubscher, (Eds.), *Invitation to psychology as a human science*. Pittsburgh, PA: Duquesne University Press.

Gordon, S. (in press b). Psycho-Neuro-Intracrinology: The mind-body continuum. In P. Snider, J. Zeff, J. Sensenig, & J. E. Pizzorno (Eds.), *The healing power of nature: The foundations of naturopathic medicine and the ecology of healing: Primary care for the twenty first century*. Philadelphia, PA: Elsevier.

Grafton, S. T., Arbib, M. A., Fadiga, L., & Rizzolatti, G. (1996). Localization of grasp representations in humans by PET 2: Observation compared with imagination. *Experimental Brain Research, 112*, 103–111.

Green, E., Green, A. M., & Walters, E. D. (1970).Voluntary control of internal states: Psychological and physiological. *Journal of Transpersonal Psychology, 2*, 1–26.

Hare, T. A., Tottenham, N., Davidson, M. C., Glover, G. H., & Casey, B. J. (2005). Contributions of amygdale and striatal activity in emotion regulation. *Biological Psychiatry, 57*(6), 624–632.

Herman, J. P., Ostrander, M. M., Mueller, N. K., & Figueiredo, H. (2005). Limbic system mechanisms of stress regulation: hypothalamo-pituitary-adrenocortical axis. *Progress in Neuropsychopharmacology and Biological Psychiatry, 29*(8), 1201–1213.

Herman, J. P., Figueiredo, H., Mueller. K., Ulrich-Lai, Y., Ostrander, M. M., & Choi, D. C.,… Cullinan, W. E. (2003). Central mechanisms of stress integration: Hierarchical circuitry controlling hypothalamo-pituitary-adrenocortical responsiveness. *Frontiers in Neuroendocrinology, 24*, 151–180.

Holland, P. C., & Gallagher, M. (1999). Amygdala circuitry in attentional and representational processes. *Trends in Cognitive Sciences, 3*(2), 65–73.

Holzbauer, M. (1976). Physiological aspects of steroids with anesthetic properties. *Medical Biology, 54*, 227–242.

Holzbauer, M., Birmingham, M. K., DeNicola, A. F., & Oliver, J. T. (1985). In vivo secretion of 3 alpha-hydroxy-5 alpha-pregnan-20-one, a potent unaesthetic steroid, by the adrenal gland of the rat. *Journal of Steroid Biochemistry, 22*(1), 97–102.

Hwang, R-J., Wu, C-H., Chen, L-F., Yeh, T-C., & Hsieh, J-C. (2009). Female menstrual phases modulate human prefrontal asymmetry: A magnoencephalographic study. *Hormones and Behavior, 55*, 203–209.

Jain, M. R., Burade, V. S., & Subhedar, N. (1995). Neurosteroids: Novel regulators of neural functions. *Indian Journal of Pharmacology, 27*, 212–221.

James, W. (1884). What is an emotion? *Mind, 9*(34), 188–205.

James, W. (1885). The function of cognition. *Mind, 10*, 27–44.

James, W. (1890). *The principles of psychology* (2 vols.). New York, NY: Henry Holt.

James, W. (1902). *The variety of religious experience: A study in human nature*. New York, NY: Longmans, Green and Co.

James, W. (1912). *The thing and its relations: Essays in radical empiricism* . (pp. 92–122). New York, NY: Longmans, Green and Co.

James, W. (1961). *Psychology briefer course*. New York, NY: Henry Holt. (Original work published 1892)

Jankord, R., & Herman, J. P. (2008). Limbic regulation of hypothalamo-pituitary-adrenocortical function during acute and chronic stress. *Annals of the New York Academy of Science, 1148*, 64–73.

Jung, C. G. (1969). *Psychology and religion: West and East* (2nd ed.) (R. F. C. Hull, Trans.). Princeton, NJ: Princeton University Press.

Jung, C. G. (1971). *On the nature of the psyche: Vol. 8, Collected works* (R. F. C. Hull, Trans.). Princeton, NJ: Princeton University Press. (Original work published 1946)

Jung. C. G. (2001). *Psychology of the unconscious: A study of the transformations and symbolisms of the libido: Collected works, Supplementary Vol. B*. (B. Hinkle, Trans.). Princeton, NJ: Princeton University Press. (Original work published 1911)

Kant, I. (2003). Transcendental doctrine of elements. In N. K. Smith (Trans.), *Critique of pure reason* (pp. 65–101). New York, NY: Palgrave Macmillan. (Original work published 1781)

Kuhn-Velten, W. N. (2000). Intracrinology and the local enzymatic control of hormone distribution and metabolism: Dehydroepiandrosterone does not just act as a prohormone for androgens and estrogens. *European Journal of Clinical Investigation, 30*(3), 34–38.

Labrie, F. (1991). Intracrinology. *Molecular Cell Endocrinology, 78*(3), C113-C118.
Labrie, F. (2003). Extragonadal synthesis of sex steroids: Intracrinology. *Annals of Endocrinology* (Paris), *64*(2), 95–107.
Labrie, F. (2004). Adrenal androgens and intracrinology. *Seminars in Reproductive Medicine, 22*(4), 299–309.
Labrie, F., Belanger, A., Cusan, L., & Candas, B. (1997). Physiological changes in dehydroepiandrosterone are not reflected by serum levels of active androgens and estrogens but of their metabolites: Intracrinology. *Journal of Clinical Endocrinology Metabolism, 82*(8), 2403–2409.
Labrie, F., Bélanger, A., Luu-The, V., Labrie, C., Simard, J., Cusan, L. et al. (1998). DHEA and the intracrine formation of androgens and estrogens in peripheral target tissues: Its role during aging. *Steroids, 63*(5–6), 322–8.
Labrie, F., Belanger, A., Simard, J., Luu-The, V., & Labrie, C. (1995). Intracrinology: Autonomy and freedom of peripheral tissues. *Annals of Endocrinology* (Paris), *56*(1), 23–29.
Labrie, F., Luu-The, V., Belanger, A., Lin, S. X., Simard, J., Pelletier, G.,…Labrie, C. (2005). Is dehydroepiandrosterone a hormone? *Journal of Endocrinology, 187*(2), 169–96.
Labrie, F., Luu-The, V., Labrie, C., & Simard, J. (2001). DHEA and its transformation into androgens and estrogens in peripheral target tissues: Intracrinology. *Frontiers in Neuroendocrinology, 22*(3), 185–212.
Langfur, S. (2012). The You-I event: On the genesis of self awareness. *Phenomenology and The Cognitive Sciences*, retrieved October 9, 2012 from http://link.springer.com/journal/11097. doi: 10.1007/s11097-012-9282-y
Legrand, D. (2007). Pre-reflective self-as-subject from experiential and empirical perspectives. *Consciousness and Cognition, 16*, 583–599.
L'Institut des Systèmes Complexes Paris Île-de-France (ISC-PIF). (Jun, 2009). Preamble. *French Complex Systems Roadmap, 2008–2009* (hal-00392486, version 1–8, pp. 5–8). Paris, France: Complex Systems Institute.
Luine, V. N., & McEwen, B. S. (1997). Effect of estradiol on turnover of the A monoamine oxidase in the brain. *Journal of Neurochemistry, 28*, 1221–1227.
MacKenzie, M. (2010). Enacting the self: Buddhist and enactive approaches to the emergence of the self. *Phenomenology and Cognitive Science, 9*(1), 75–99.
MacKenzie, M. (2012). Enacting the self: Buddhist and enactive approaches to the emergence of the self. In M. Sidertis, E. Thompson, and D. Zahavi (Eds.), *Self, no self? Perspectives from analytical phenomenological, & Indian traditions* (pp. 239–273). New York, NY: Oxford University Press. (Reprinted from 2010, *Phenomenology and the Cognitive Sciences, 9*, 75–99.)
Majewska, M. D. (1987). Steroids and brain activity: Essential dialogue between body and mind. *Biochemical Pharmacology, 36*, 3781–3788.
Majewska, M. D., Harrison, N. L., Schwartz, R. D., Barker, J. L., & Paul, S. M. (1986). Steroid hormone metabolites are barbiturate-like modulators of the GABA receptor. *Science, 232*, 1004–1007.
Maslow, A. (1954). *Motivation and personality*. New York, NY: Harper and Row.
Maslow, A. (1971). A theory of metamotivation: The biological rooting of the value life. In A. Maslow (Ed.), *Farther reaches of human nature* (pp. 299–339). New York, NY: Viking. (Original work published 1967)
Maturana, H. R. (1975). The neuropsychology of cognition. In P. Garvin (Ed.), *Cognition: A multiple view*. New York, NY: Spartan.
Maturana, H. R. (1988a). Reality: The search for objectivity or the quest for a compelling argument. *Irish Journal of Psychology, 9*, 25–82.
Maturana, H. R. (1988b). Ontology of observing: The biological foundations of self, consciousness and the physical domain of existence. In R. Donaldson (Ed.), *Texts in cybernetic theory: An in-depth exploration of the thought of Humberto Maturana, William T. Powers, and Ernst von Glaserfeld*. Felton, CA: American Society for Cybernetics. Retrieved 9/4/12 from http://www.inteco.sl/biology/ontology/index.htm.

Maturana, H. R., & Varela, F. J. (1980). *Autopoiesis and cognition: The realization of the living.* Boston, MA: D. Reigel.

Maturana, H. R., & Varela, F. J. (1987). *The tree of knowledge: The biological roots of human understanding*. Boston, MA: Shambala.

Maturana, H. R., Varela, F. J., & Uribe, R. (1974). Autopoiesis: The organization of living systems, its characterization and a model. *Biosystems, 5*, 187–196.

May, R. (1975). *The courage to create*. New York, NY: Norton.

May, R. R., Angel, E., & Ellenberger, H. F. (1958). *Existence: A new dimension in psychiatry and psychology*. New York, NY: Basic Books.

Melzoff, A., & Moore, M. K. (1977). Imitation of facial and manual gestures by human neonates. *Science, 198*, 75–78.

Melzoff, A., & Moore, M. K. (1983). Newborn infants imitate adult facial gestures. *Child Development, 54*, 702–709.

Melzoff, A., & Moore, M. K. (1989). Imitation in newborn infants: Exploring the range of gestures imitated and the underlying mechanisms. *Developmental Psychology, 25*, 954–962.

Merleau-Ponty, M. (1962). *Phenomenology of perception*. New York, NY: Routledge & Kegan Paul. (Original work published 1945)

Myers, F. W. H. (1886). Multiplex personality. *Proceedings of the English Society for Psychical research, 4*, 496–514.

Myers, F. W. H. (1892).The subliminal consciousness: Chapter 1. General characteristics of subliminal messages. *Proceedings of the Society for Psychical Research, 7*, 298–327.

Myers, F. W. H. (2001). *Human personality and its survival of bodily death*. Charlottesville, VA: Hampton Roads. (Original work published 1903)

Noë, A. (2004). *Action in perception*. Cambridge, MA: The MIT Press.

Ochsner, K. N., Bunge, S. A., Gross, J. J., & Gabrieli, J. D. (2002). Rethinking feelings: An fMRI study of the cognitive regulation of emotion. *Journal of Cognitive Neuroscience 14*(8), 1215–1229.

Ochsner, K. N., Ray, R. D., Cooper, J. C., Robertson, E. R., Chopra, S., & Gabrieli, J. D. E,... Gross, J. J. (2004). For better or worse: Neural system supporting the cognitive down-and up-regulation of negative emotion. *NeuroImage, 23*(2), 483–499.

Ochsner, K. N., Ray. R. R., Hughes, B., McCrae, K., Cooper, J. C., Weber, J.,...Gross, J. J. (2009). Bottom up and top down processes in emotion generation: Common and distinct central mechanism. *Psychological Science, 20*(11), 1322–13331.

Papez, J. W. (1937). A proposed mechanism of emotion. *Archives of Neurological Psychiatry, 38*,725-743.

Pert, C. B. (1999). *Molecules of emotion*. New York, NY: Touchstone.

Phelps, E. A. (2006). Emotion and cognition: Insights from studies of the human amygdala. *Annual review of Psychology, 57*, 27–53.

Phelps, E. A., Ling, S., & Carrasco, M. (2006). Emotion facilitates perception and potentiate the perceptual benefits of attention. *Psychological Science, 17*(4), 292–299.

Radley, J. J., Arias, C. M., & Sawchenko, P. E. (2006). Regional differentiation of the medial prefrontal cortex in regulating adaptive responses to acute emotional stress. *The Journal of Neuroscience, 26*(50), 12967–12976.

Rapkin, A. J. (1999). Progesterone, GABA and mood disorders in women. *Archives of Women's Health, 2*, 97–105.

Re, R. N. (2003). The intracrine hypothesis and intracellular peptide hormone action. *BioEssays, 25*, 401–409.

Re, R. N. (2002). The origins of intracrine hormone action. *American Journal of Medical Science, 323*(1), 43–48.

Re, R. N., & Bryan, S. E. (1984). Functional intracellular rennin-angiotensin systems may exist in multiple tissues. *Hypertension Clinical Experimental Theory Tract, A6*(Suppl. 10 & 11), 1739–1742.

Re, R. N., & Cook, J. L. (2008). The basis of an intracrine pharmacology. *Journal of Clinical Pharmacology, 48*(3), 344–350.

Re, R. N., & Cook, J. L. (2006). The intracrine hypothesis: An update. *Regulatory Peptide, 133*(1–3), 1–9.
Risold, P. Y. & Swanson, L. W. (1996). Structural evidence for functional domains in the rat hippocampus. *Science, 272*, 1484–1486.
Rizzolatti, G., Fadiga, L., Gallese, V., & Fogassi, L. (1996). Premotor cortex and the recognition of motor actions. *Cognitive Brain Research, 3,* 131–141.
Robel, P., & Baulieu, E. E. (1995). Dehydroepiandrosterone (DHEA) is a neuroactive steroid. *Annals of the New York Academy of Science, 774*, 82–110.
Rogers, C. (1964). Toward a science of the person. In T. W. Wann (Ed.), *Behaviorism and phenomenology: Contrasting bases for modern psychology* (pp. 109–140). Chicago, IL: University of Chicago Press.
Rupprecht, R., Koch, M., Montkowski, A., Lancel, M., Faulhaber, J., & Harting, J.,…Spanagel, R. (1999). Assessment of neuroleptic-like properties of progesterone *Psychopharmacology, 143*(1), 29–38.
Ryan, J., & Herman, H. P. (2008). Limbic regulation of hypothalamo-pituitary-adrenocortical function during acute and chronic stress. *Annals of the New York Academy of Science, 1148*, 64–73.
Sanders, D., Warner, P., Backstrom, T., & Bancroft, J. (1983). Mood, sexuality, hormones and the menstrual cycle: Change in mood and physical state: Description of subjects and method *Psychosomatic Medicine, 45*(6), 487–501.
Sansonese, J. N. (1994). *The body of myth: Mythology, shamanic trace, and the sacred geography of the body*. Rochester, VT: Inner Traditions International.
Searle, J. (2005). The self as a problem in philosophy and neurobiology. In T. E. Feinberg & J. P. Keenan (Eds.), *The lost self: Pathologies of the brain and identity*. Oxford, NY: Oxford University Press.
Selye, H. (1946). Correlations between the chemical structure and the pharmacologic actions of steroids. *Endocrinology, 30*, 437–453.
Sherwin, B. B., & Surayani-Cadotte, B. E. (1990). Up-regulatory effect of estrogen on platelet 3H-impramine binding sites in surgically menopausal women. *Biological Psychiatry, 28*, 339–348.
Siderits, M., Thompson, E., & Zahavi, D. (Eds.). (2011). *Self, no self? Perspectives from analytical, phenomenological, and Indian traditions*. New York, NY: Oxford University Press.
Simpson, E. R. (2002). Aromatization of androgens in women: Current concept and findings. *Fertility and Sterility, 77*(4), S6-S10.
Squire, L. R., Berg, D., Bloom, F. E., Du lac, S., Ghosh, A., & Spitzer N. C. (2008). *Fundamental neuroscience* (3rd ed.). Burlington, MA: Elsevier Inc.
Strauss, J. F., & Barbieri, R. L. (Eds.). (2009). *Yen and Jaffe's reproductive endocrinology: Physiology, pathophysiology, and clinical management* (6th ed.). Philadelphia, PA: W. B. Saunders.
Taylor, E. I. (1984). *William James on exceptional mental states: The 1896 Lowell Lectures*. Amherst, MA: The University of Massachusetts Press.
Taylor, E. I. (1996). *William James on consciousness beyond the margin*. Princeton, NJ: Princeton University Press.
Taylor, E. I. (2009). *The mystery of personality: A history of psychodynamic theories*. New York, NY: Springer.
Thompson, E. (2005). Sensorimotor subjectivity and the enactive approach to experience. *Phenomenology and the Cognitive Sciences, 4*(4), 407–427.
Thompson, E. (2007). *Mind in life: Biology, phenomenology, and the sciences of mind*. Cambridge, MA: Belknap Press of Harvard University.
Thompson, E., & Varela, F. J. (2001). Radical embodiment. *Trends in Cognitive Science 5*(10), 418–425.
Thompson, J., Maddock, M., Ayaland, M., & Oswald, I. (1977). Relationship between nocturnal plasma oestrogen concentration and free plasma tryptophan in perimenopausal women. *Journal of Endocrinology, 72*, 395–396.

Tsigos, C., & Chrousos, G. P. (2002). Hypothalamic-pituitary-adrenal axis, neuroendocrine factors and stress. *Journal of Psychosomatic Research, 53*(4), 865–871.
Varela, F. J. (1979a). *Principles of biological autonomy.* New York, NY: Elsevier/North Holland.
Varela, F. J. (1981). Describing the logic of the living. In M. Zeleny (Eds.), *Autopoiesis: A theory of living organization* (pp. 36–48). New York, NY: Elsevier.
Varela, F. J. (1991). Organism: A meshwork of selfless selves. In A. I. Tauber (Ed.), *Organism and the origins of self* (pp. 79–107). Netherlands: Kluwer Academic Publishers.
Varela, F. J. (1992). *Ethnical know-how: Action, wisdom and cognition.* Stanford, CA: Stanford University Press.
Varela, F. J. (1999). The specious present: A neurophenomenology of time consciousness. In J. Petitot, F. J. Varela., B. Pachoud, & J-M. Roy (Eds.), *Naturalizing phenomenology: Issues in contemporary phenomenology and cognitive science* (pp. 266–314). Stanford, CA: Stanford University Press.
Varela, F. J., & Cohen, A. (1989). Le corps evocateur: Une relectre de L'immunité. *Novelle Revue de Psychanalyse, 40*, 193–213.
Varela, F. J., & Goguen, J. (1977). The arithmetic of closure. In R. Trappel (Ed.), *Progress in cybernetics and systems research* (Vol. 3, pp. 48–63). New York, NY: Wiley Hemisphere.
Varela, F. J., Thompson, E., & Rosch, E. (1991). *The embodied mind: Cognitive science and human experience*. Cambridge, MA: The MIT Press.
Vaz, N. M., & Varela, F. J. (1978). Self and non-sense: An organism-centered approach to immunology. *Medical Hypothesis, 4*(3), 231–261; 263–267.
Vuilleumier, P. (2009). The role of the human amygdala in perception and attention. In P. J. Whalen & E. A. Phelps (Eds.), *The human amygdala* (pp. 220–249). New York, NY: The Guilford Press.
Watts, A. (1975). *Psychotherapy east and west.* New York, NY: Vintage Books.
Whalen, P. J., Kagan, J., Cook, R. G., Davis, F. C., Kim, H, & Polis, S.,…Johnstone, T. (2004). Human amygdala responsivity to masked fearful eye whites. *Science, 306*, 2061.
Wiener, N. (1948). *Cybernetics of control and communication in the animal and in the machine.* New York, NY: Wiley.
Zahavi, D., & Roepstorff, A. (2011). Faces and ascriptions: Mapping measures of the self. *Consciousness and Cognition, 20*, 141–148.
Zahavi, D. (2008). *Subjectivity and selfhood: Investigating the first-person perspective*. Cambridge, MA: The MIT Press.

Glossary

Autopoiesis (self-producing) Refers to the self-creating, self-organizing properties of chemical, neuronal, and cognitive systems through which they continually regenerate reproducing themselves by their own mutual interactions. The theory of autopoiesis defines the uniqueness of the emergence that produces life in its fundamental cellular form. For example, the living cell, when conceptualized as an autopoietic system is the continual creation of itself. Biochemical pathways of the cell and its membranes continuously regenerate through the internal production of substratum components, and biochemical states of the organism transform the state of activity of neural networks by acting on the neuron's membrane receptors.

Befindlichkeit A term used by Heidegger. The concept is sometimes translated as "state of mind" but is closely related to the everyday concepts of mood or emotion. As Gendlin points out, the translation of *befindlichkeit* into "state of mind" does not quite capture Heidegger's original meaning. The German phrase *sich befinden* translates as "finding oneself." It is perhaps best translated into English as "how-you-are-ness," or "self-finding."

Cognitivism A philosophy of the mind, which proposes that perceiving, learning, and thinking can be explained as an unproblematic and mechanical reflective process governed by symbolic manipulation of representations in the mind as well as preprogrammed procedural operations.

Consciousness Waking rational consciousness, in which the person is aware only of what is at the center of the field of attention; also used to refer to the view that our experience is of a field with a focus and a margin, where the margin keeps changing – expanding and contracting in the form of a stream that flows ever onward; also used to refer to all the states of consciousness it is possible for a person to experience, ranging from the psychopathic, through the normal, to the transcendent.

Critical constructivism In education this concept stresses the active creation of learning that the student contributes to; critical constructivism avoids privileging powerful and pervasive ideologies influencing education and includes the role of the teacher as co-constructing the value of what is being learned.

S. Gordon (Ed.), *Neurophenomenology and Its Applications to Psychology*,
DOI 10.1007/978-1-4614-7239-1,

Deity yoga A type of Tibetan meditation in which the meditator focuses their attention on an internally generated image of the deity and identifies with it.

Dynamic theories of the unconscious The attempt to study altered states of consciousness from the standpoint of the waking rational state alone; also used in the sense of the transmutation of consciousness from one state to another, such as the transition between waking and dreaming and back again.

Eidetic reduction The second step in Edmund Husserl's philosophical method of phenomenology, the first step being the phenomenological reduction. Within the eidetic reduction, the investigator uses careful description of phenomenon as they are given to awareness (within the phenomenological reduction) and then uses imaginative variation of profiles of a phenomenon in order to eliminate nonessential characteristics of the phenomenon. The aim is to arrive at only those profiles of the phenomenon that are essential to its structure.

Embodiment Refers to the bodily aspects of human subjectivity and the biological and physical presence of our body as a necessary precondition for the experience of emotion, language, thought, and social interaction. It is our kinesthetic awareness of the body as the vehicle through which we experience the sensorimotor, perceptual, and nonconceptual lived world. This is not a cognitive understanding of self in the world, but a proprioceptive, nonconceptual awareness that is tacit, pre-reflective, and intersubjective. In relation to cognitive neuroscience and neurophenomenology, having an affective, sensing body is integral to human understanding.

Empirical phenomenology An approach to psychology and the social sciences founded on Husserlian phenomenology to develop a qualitative analysis of a participant's experience. The aim of the research is to describe in rich, thick language the experience of participants, and through the use of imaginative variation and other procedures, to perform a qualitative analysis, which identifies the meaning of the experience for the participants.

Enactive Jerome Bruner (1966) first used this term to explain his theory of how children learn using three successive modes of representation: enactive (action-based), iconic (image-based), and symbolic (language-based). The phrase, "enactive approach to cognition," which was subsequently used by Varela, Thompson, and Rosch (1991) referred to embodied action in lived experience. Living beings are autonomous agents that actively generate and maintain their identities and *enact* or bring forth their own cognitive domains. The nervous system, as an autonomous system, generates and maintains its own coherent and meaningful patterns of activity according to its operation as an organizationally closed or circular and reentrant sensorimotor network of interacting neurons. The nervous system does not process information in the computationalist sense but creates meaning. Cognition is a form of embodied action (cognitive structures emerge from recurrent sensorimotor patterns of perception and action). A cognitive being's world is not a prespecified, external realm, represented internally by its brain, but a relational domain that is enacted by that being's autonomous agency and mode of coupling with the environment.

Enactivism The theory that human knowing and purposeful acting is best understood as a complex interrelationship between brain, body, and environment. Enactivism proposes that knowledge emerges from dynamic and adaptive bodily action with and in the environment.

"Existential-phenomenological state" A term developed by Susan Gordon (Chap. 5) to describe the present state of the organism in constant flux; the term incorporates biopsychosocial, spiritual, transcendent, and subliminal aspects of the person in "sense of self." The focus on the "self" is not on its measurement, but on its direct experience through the actualization of its potential and its striving towards health as intrinsic to human motivation.

"Experiential revolution" A term developed by Brent Dean Robbins (Chap. 1) to describe the latest in a series of "revolutions" in psychology. The "cognitive revolution" leads to an advance on behaviorism by permitting the investigation of "internal" processes of information processing and perception and relied primarily upon a computer metaphor of the mind. The "affective revolution" was the period in the 1990s which saw an increasing focus upon the study of affect (moods and emotions) in addition to computerlike information processing. The "experiential revolution" is the latest trend in the social, behavioral, and biological sciences in which there is an increasing focus on understanding human experience on its own terms with fewer tendencies to approach experience by reducing it to biology (see "Reductionism").

HPG-HPA axes Refers to the hypothalamic-pituitary-gonadal (HPG) axis and hypothalamic-pituitary-adrenal (HPA) axis, which functionally compete for hormonal substrate to balance reproductive and adrenal function by up- and down-regulating DHEA and progesterone to their bioactive metabolites, which are immediately necessary for the individual's homeostasis.

Integrative phenomenology The integration of philosophical and empirical phenomenology with qualitative and quantitative evidence in psychology and other behavioral and social sciences.

Intracrine Refers to peptide/protein hormones that have intracellular actions and/or act within the cell that synthesizes it, i.e., hormones that are synthesized and exert their action in the same cell without exit and reentry (in contrast to autocrine signaling where an effector molecule leaves the cell of its origin before it rebinds to the surface of that cell).

Mandala meditation A type of Tibetan meditation in which the meditator focuses attention on the internally generated image of the deity with his or her habitat, but continues looking at the picture instead of becoming a deity as in Deity Yoga.

Meditation A volitional change in mental functioning that usually includes slowing down of thoughts, awareness or witnessing the contents of the mind, introspection, and interruption of uncontrolled, random activities of the mind with the increase of spaciousness and presence. Meditation always involves physical relaxation. Various types of meditation may include generation of imagery or thoughts with spiritual or religious contents, focusing attention on breath or sound, and other specific modalities of mental activity. It is traditional to distinguish between types of meditation in which awareness predominates and focusing predominates, even though both heightened awareness and focusing of the mind are frequently combined. Recently, distinctions were introduced between non-dual and dualistic meditation or between meditation where attention is focused on something (referential mode) and where attention is diffuse (non-referential mode). Classification of the types of meditation is not completed because of the complexity and diversity of the mental functions that can be modified in meditation. Traditionally, meditation is used in spiritual or religious systems as an exercise that changes the mind towards an established transformational soteriological goal such as enlightenment. In scientific research, meditation is used to decrease stress and to enhance cognitive functioning.

Meditation styles The term introduced by scientists in order to account for the difference in the effects of the different kinds of meditation.

Mythopoesis A word used in the late nineteenth century by the English classicist Fredrick William Henry Myers to indicate a dynamic psychological relation between myths and poetic language. Essentially this means that poetry is a language that often grasps at the inexpressible, and in good poetry, mythopoetic language succeeds in elevating personal experience to a level where an individual's reality is believed to unite with the experience of all humanity.

Neurocorrelates of experience Functional or anatomical changes in the brain associated with certain types of experience or the parts of experience.

"Neurophenomenological self" A term developed by Susan Gordon (Chap. 5) to describe the growth-oriented, self-actualizing dimension of the person. This construct depicts an enactive, embodied "sense of self" rather than its mental representation.

Neurophenomenology A movement in the neurosciences to establish a dialogue between first-person science, usually in terms of Husserlian phenomenology, and third-person science, i.e., the cognitive neurosciences, redefining the cognitivists' agenda by expanding the definition of cognition beyond the principles of S-R theory to include a cognition experienced in the context of a physical body and an environment.

Neurophilosophy An approach to the philosophy of mind that is most identified with the work of Patricia and Paul Churchland. The aim of neurophilosophy has been to explain psychological constructs and mental phenomenon according to a biologically reductionistic model.

No-self Refers to the Buddhist doctrine of nonself or *anātman*, a rejection of the atmān or the enduring substantial self as a type of self-reference. Theorists do not infer from the sense of self that self really exists because they claim that it is not ontologically grounded, but arises from the stream of consciousness.

Phenomenological-cognitive mapping A term and procedure developed by Olga Louchakova Schwartz (Chap. 3) to describe a research procedure by which the information concerning the phenomenological structures of experience is "translated" into the language of the cognitive scientific experiment. Since the cognitive variables are, in fact, the modalities of experience identified in the natural attitude, such a procedure seems to be redundant. However, in dealing with the high complexity, fluidity, and empirical uniqueness of individual experiences, such a procedure allows to standardize the forms of experience that can be juxtaposed and successfully tested in the cognitive experiment.

Phenomenological reduction The first step in Edmund Husserl's philosophical method of phenomenology, the second step being the eidetic reduction. Within the phenomenological reduction, the investigator sets aside the convention distinction between subjectivity and objectivity and brackets these presuppositions in order to observe phenomenon as they are actually given to experience prior to such metaphysical categories.

Phenomenology A philosophical tradition dating from the early nineteenth century emanating from European sources and claimed to be the exclusive province of Edmund Husserl and his followers; alternately, a focus on lived experience in the immediate moment.

Phenomenology is the study of phenomena, the way things appear to us in experience or consciousness. Kant used the term to distinguish between the study of objects and events as they appear in experience (phenomena) and as they are in themselves (noumena). Hegel used the term to describe the science in which we come to know mind as it is in itself and as it appears to us. Husserl's transcendental or constitutive phenomenology became a method of self-discovery and ego transcendence, a human science movement, and a discipline endeavoring to describe how the lifeworld is constituted and experienced by reducing or bracketing distorting influences of various cognitive operations upon the object. To Heidegger, phenomenology was ontology, the study of modes of being in the world. To Merleau-Ponty, it was the study of essences, the nature or meaning of perception. To James, phenomenology was radical empiricism, lived experience in the immediate moment before the differentiation of subject and object.

Phenomenology of the science-making process itself A phrase developed by Eugene Taylor (Chap. 4). While neuroscience represents the study of the biology of consciousness, certain humanistic implications are generated from this study that suggest the key to understanding the hard problem – that is, the relation between understanding a thing and the immediate experience of it – may be found within the phenomenological lifeworld of the experimenter.

Portfolio assessment A cognitive-developmental and pedagogical praxis used by teachers in which the teacher methodically gathers a wide variety of student learning accomplishments (e.g., observations, student reflections, student chosen work, quantitative and qualitative assessments) in order to form a holistic assessment of the unique learner. Importantly, the portfolio assessment spans, at least, the entire class semester of student work.

Pragmatism An American philosophical movement initiated by Charles Sanders Peirce and William James, who could not agree on a single definition of the term so there developed "the two pragmatisms." One was the Peircean view, where, as a rule of logic, in order to be clear, one needs to consider the consequences of one's claims. The other was the Jamesean view: that beliefs are tested by their consequences. The Peircean view became the foundation of analytic philosophy in the twentieth century, while the Jamesean view dominated the progressive era of American popular philosophy in the opening decades of the twentieth century. Pragmatism was also a method for evaluating competing truth claims, where many competing and apparently contradictory philosophies could be found to lead to the same ends, functionally suggesting that they were equal, but not the same.

Praxis Denotes when research informs practical activity and pedagogical exercises. In relation to critical pedagogy, praxis is metacognitive in regard to practical application, which, in turn, leads to critique and reform. Praxis leads to increased learning agency when new forms of learning are discovered and old forms of learning, as well as particular knowledge claims, may be deemed oppressive.

Preflective Refers to preconceptual or pre-languaged bodily knowing, which exists prior to felt sense and allows for meaningful experience on the level of conceptual awareness. It is the awareness we have before we do any reflecting on our experience; an implicit and first-order awareness rather than an explicit or higher-order form of self-consciousness. An explicit reflective self-consciousness is possible only because there is a pre-reflective self-awareness that is an ongoing and more primary self-consciousness. The experiential dimension always involves implicit pre-reflective self-awareness. Husserl maintained that consciousness involves self-appearance; Merleau-Ponty believed consciousness is given to itself and has no meaning independently of this self-givenness; Sartre thought pre-reflective self-consciousness constitutes the mode of being of the experience. Its subjective "feel" or phenomenal quality; "what it is like" or what it "feels" like to have experiences.

Pluralism Or more properly noetic pluralism, referring to the independent nature of visionary consciousness experienced within each individual, in all its vicissitudes, including the ability to experience the transcendent consciousness of oneness, except that this oneness might be different from person to person.

Psychology as epistemology The idea that from a phenomenological standpoint there is no such thing as a science without the person, though contemporary science insists with its hyperobjectivist stance that the experimenter has no effect on the outcome of the experiment.

Psycho-neuro-intracrinology A construct developed by Susan Gordon (Chap. 5) that combines three complex systems, which are impossible to tease apart; represents the embodied self; examines the interrelationship between psychological, neurological, and intracrinological processes forming a mind-brain continuum within the person. Psycho (psychological) refers to constructs variously referred to as psyche, self, soul, mind, and consciousness. Neuro (neurological) refers to the composition and reactions within the nervous system. Intracrine (intracrinological) refers to the micro-level synthesis and regulation of hormones and their enzymes that modulate feedback and feedforward from upstream substrates to determine genetic self-expression. Intracrinology is the study of the intracellular biosynthesis of steroids, the binding of receptors, and the formation of enzymes that catalyze the creation of hormones within the cell.

Radical empiricism Pure experience before the differentiation between subject and object.

Reductionism An approach to science that assumes that all complex systems in social, behavioral, and biological sciences can be reduced to and explained by more basic elemental causal forces. Reductionism is perhaps epitomized by Francis Crick's argument that the aim of modern biology is to explain biology in terms of physics and chemistry. An extreme version of this argument is the view held by neuroscientists that all mental states are only chemical reactions in the brain.

Rig-pa A non-referential mode of meditation, open awareness, which is experienced as distinct and real, as opposed to the contents of the mind, which appears not real against the backdrop of open space-like awareness. Rig-pa is a developmental condition, with different stages.

Sense of self Refers to pre-reflective, autonomic, subliminal experience; embodiment as lived from the point of view of the subject's first-person experience associated with cognitive and mental events (attention, present-time consciousness, body image, volition, perceptual filling in, fringe, center, and emotion), which represent an irreducible ontological level that retains its quality of immediacy because it plays a role in the organism's structural coherence.

Situated cognition Proposes that human perceiving, knowing, learning, and thinking are best understood in relation to specific and concrete social situations that can be systematically observed.

Subjugation In terms of learning, this term denotes the educational disciplining of the learner as a means of control and hegemony. The learner may be said to be subjugated when overly restrictive and predictive pedagogies serve to reinforce positivist education paradigms alone and thereby misrepresent others' ways of learning, knowing, and thinking. Subjugation in education leads to passive educable subjects.

Tibetan meditation A generic name for the practice within the Vajrayana branch of Buddhism developed in Tibet. Tibetan Buddhism involves an enormously broad spectrum of different kinds of meditation, as in fact every kind of mental activity can be turned into a means of meditation.

Transcendental phenomenology Edmund Husserl's approach to phenomenology in which the phenomenological reduction is preparation for the eidetic reduction. The aim of transcendental phenomenology is to identify the essence, or *eidos*, of a given phenomenon (see "Eidetic Reduction").

Transformative learning Highlights what the adult learner interpretively brings to any given learning situation in order to transform previously acquired knowledge paradigms.

Vipashyana A meditative practice of witnessing different elements of the mind, a discerning state of mind, as opposed to the regular state in which mind functions as one with its contents.

About the Editor and Authors

About the Editor

Susan Gordon, Ph.D. is Core Adjunct Professor of Psychology at National University, La Jolla, CA, and Research Director of the Southbury Clinic for Traditional Medicines, Southbury, CT. She has a doctorate in the History and Philosophy of Psychology (mind-body medicine) from Saybrook University, has completed coursework in naturopathic medicine at Bastyr University, and is affiliated with Harvard College Library, Institut des Systèmes Complexes, Paris Île-de-France, and Bibliothèque François Mitterrand. Her research and publications integrate holistic health, cognitive neuroscience, reproductive endocrinology, and existential-phenomenology. She serves on the Executive Board of the Society for Humanistic Psychology, Division 32 of the American Psychological Association, and the editorial review boards of *The Humanistic Psychologist*, *Philosophy Study*, and *PsycCRITIQUES*. She authored "Psycho-Neuro-Intracrinology: The mind-body continuum" in *The Healing Power of Nature: The Foundations of Naturopathic Medicine and the Ecology of Healing: Primary Care for the Twenty First Century* (Elsevier, in press) and is contributing author to the forthcoming *Invitation to Psychology as a Human Science* (Duquesne University Press) and *The Handbook of Humanistic Psychology* (Sage, 2nd ed.).

S. Gordon (Ed.), *Neurophenomenology and Its Applications to Psychology*,
DOI 10.1007/978-1-4614-7239-1, © Springer Science+Business Media New York 2013

About the Authors

Brent Dean Robbins, Ph.D. is Associate Professor of Psychology at Point Park University, Pittsburgh, PA, and Director of the Psychology Program. He has a doctorate in Clinical Psychology from Duquesne University and is an outpatient therapist at Mercy Behavioral Health. He is Editor-in-Chief of *Janus Head: Journal of Interdisciplinary Studies*, which examines the intersections among Continental philosophy, the social sciences, literature, and the arts. He is President-Elect of the Society for Humanistic Psychology and serves on the editorial boards of the *Journal of Humanistic Psychology, The Humanistic Psychologist, International Journal of Transpersonal Studies, PsycCRITIQUES, International Journal of Existential Psychology and Psychotherapy*, and the *Indo-Pacific Journal of Phenomenology*. His publications include studies of the socioeconomical conditions that influence the diagnosis and treatment of mental health, the phenomenology of emotion and embodiment, sociocultural approaches to existential therapy, and neurophenomenology. He is coeditor of *Drugging Our Children: How Profiteers Are Pushing Antipsychotics on Our Children and What We Can Do to Stop It* (Praeger, 2012) and *The Legacy of R.D. Laing* (Trivium, 2012).

Robert G. McInerney, Ph.D. is Associate Professor of Psychology at Point Park University, Pittsburgh, PA. He has a doctorate in Clinical Psychology from Duquesne University and certification in qualitative human science research from the Center for Interpretive and Qualitative Research at Duquesne. He cochaired the Fifth Annual Conference of the Society for Humanistic Psychology and is on the editorial review board of the *Journal of Humanistic Psychology*. His research interests include hermeneutic phenomenology, social constructionism, and community psychology. Relevant publications include "Oz Never Did Give Nothing to the… Scarecrow: Neurophenomenology and Critical Pedagogy" in *Phenomenology and Practice* (2010) and the coauthored "Using Neuropsychology to Enhance Existential Psychotherapy" in *The Humanistic Psychologist* (forthcoming) and "Toward a Method of Neurophenomenological Assessment and Intervention" in *The Humanistic Psychologist* (2002). He is currently working with ASSET (Achieving Student Success through Excellence in Teaching) to provide pedagogical applications of metacognition and enactive learning for STEM (science, technology, engineering, and math) primary-school educators.

Olga Louchakova-Schwartz, M.D., Ph.D. is Professor of Psychology and Comparative Religion at Sofia University and Founding Director of the Neurophenomenology Center, Palo Alto, CA. She received her M.D. from Leningrad Pediatric University and Ph.D. from the Academy of Medical Sciences, Russia. A recognized spiritual teacher, she founded Hridayam® School of Psychospiritual Development. Her chief research contributions are in transdisciplinary research methods, such as the early diagnostics of demyelination or genetic phenomenological analysis applied to Vedanta and Hesychasm, embodiment and the psychology of spiritual development. Her collaborative research of visual capacities of Tibetan monks was featured on *Science Daily*. Her numerous publications cover a broad range of topics from the autoimmune diseases of the nervous system to psychosomatic mysticism, intrapsychic ontopoiesis, and nondual consciousness. Her past appointments include Senior Scientist at the Pavlov Institute of the Russian Academy of Sciences and Visiting Professor in the Graduate Theological Union in Berkeley and California Institute of Integral Studies in San Francisco. She is presently working on comparisons between phenomenology and Islamic metaphysics.

Eugene Taylor, Ph.D. (1946–2013) was a beloved Professor in the School of Psychology and Interdisciplinary Inquiry at Saybrook University, San Francisco, CA; Lecturer on Psychiatry at Harvard Medical School, Boston, MA; and Senior Psychologist on the Psychiatry Service at the Massachusetts General Hospital. He holds a doctorate in the History and Philosophy of Psychology from the University Professors Program at Boston University. He is an internationally renowned scholar on the life and work of William James, while he sees his own primary contribution to psychology as the development of a phenomenologically oriented intuitive psychology of spiritual self-realization. He has authored 35 years worth of journal articles, book chapters, and book reviews. His book-length publications include *Stanley Cobb: Builder of the Modern Neurosciences* (1984), with Benjamin White and Richard Wolfe; *William James on Exceptional Mental States* (1982); *William James on Consciousness beyond the Margin* (1996); *Pure Experience: The Response to William James* (1996), with R. Wozniak (Eds.); *The Mystery of Personality: A History of Psychodynamic Theories* (2009); and *William James and the Spiritual Roots of American Pragmatism* (forthcoming).

Bibliography

Abraham, J. (2003). Dynamic systems theory: Applications to pedagogy. In T. Schacher & A-P., Dauwalder (Eds.), *The dynamical systems approach to cognition: Concepts and empirical paradigms based on self-organization, embodiment and coordination dynamics* (pp. 295–308). Singapore: World Scientific Publishing Company.

Abramowitz, E. S., Bakerk, A. H., & Fleisher, S. F. (1982). Onset of depressive psychiatric crisis and the menstrual cycle. *American Journal of Psychiatry, 139,* 475–478.

Adams, F. (2010). Embodied cognition. *Phenomenology and Cognitive Science, 9,* 619–628.

Adolphs, R., Gosselin, F., Buchanan, T. W., Tranel, D., Schyns, P., & Damasio, A. R. (2005). A mechanism for impaired fear recognition after amygdala damage. *Nature, 433,* 68–72.

Adolphs, R., Tranel, D., Hamann, S., Young, A.W., Calder, A. J., Phelps, E. A.,…Damasio, A. R. (1999). Recognition of facial emotion in nine individuals with bilateral amygdala damage. *Neuropsychologia, 37,* 1111–1117.

Aftanas, L. I., & Golocheikine, S. A. (2001). Human anterior and frontal midline theta and lower alpha reflect emotionally positive state and internalized attention: High resolution EEG investigation of meditation. *Neuroscience Letters, 310*(1), 57–60.

Aftanas, L. I., & Golocheikine, S. A. (2002). Non-linear dynamic complexity of the human EEG during meditation. *Neuroscience Letters, 330,* 143–146.

Ales Bello, A. (2009). The divine in Husserl and other explorations. *Analecta Husserliana: Vol. 98*. New York, NY: Springer.

Amaral, D. G., Behniea, H., & Kelly, L. (2003). Topographic organization of projections from the amygdala to the visual cortex in the Macque monkey. *Neuroscience, 118*(4), 1099–1200.

Anderson, A. K., & Phelps, E. A. (2001). Lesions of the human amygdala impair enhanced perception of emotionally salient events. *Nature, 411,* 305–309.

Anderson, A. K., & Phelps, E. A. (2002). Is the human amygdala critical for the subjective experience of emotion? Evidence of intact dispositional affect in patients with amygdala lesions. *Journal of Cognitive Neuroscience, 14*(5), 709–720.

Armony, J. L., Servan-Schreiber, D., Cohen, J. D., & LeDoux, J. E. (1997). Computational modeling of emotion: Explorations through the anatomy and physiology of fear conditioning. *Trends in Cognitive Science, 1,* 28–34.

Aukrust, V. G. (Ed.). (2011). *Learning and cognition in education.* London, UK: Elsevier Academic Press.

Baars, B. J. (1986). *The cognitive revolution in psychology.* New York, NY: The Guilford Press.

Baars, B. J. (1999) Attention vs. consciousness in the visual brain: Differences in conception, behavior, neuroanatomy, and physiology. *Journal of General Psychology, 126*(3), 224–233.

Bader, J. (1990). Meditation in Shankara's Vedanta. New Delhi, India: Aditya Prakashan.

S. Gordon (Ed.), *Neurophenomenology and Its Applications to Psychology*,
DOI 10.1007/978-1-4614-7239-1, © Springer Science+Business Media New York 2013

Baerentsen, K. B., Stødkilde-Jørgensen, H., Sommerlund, B., Hartmann, T., Damsgaard-Madsen, J.,…Green, A. C. (2010). An investigation of brain processes supporting meditation. *Cognitive Processing, 11*(1), 57–84.

Baker, D. B. (Ed.). (2012). *The Oxford handbook of the history of psychology: Global perspectives*. Oxford, NY: Oxford University Press.

Banissy, M. J., Kanai, R., Walsh, V., & Rees, G. (2012). Inter-individual differences in empathy are reflected in human brain structure. *Neuroimage, 62*(3), 2034–2039.

Banquet, J. P., & Lesèvre, N. (1980). Event-related potentials in altered states of consciousness. *Progress in Brain Research, 54,* 447–453.

Barclay, M. W. (2000). The inadvertent emergence of a phenomenological perspective in the philosophy of cognitive psychology and psychoanalytic developmental psychology. *Journal of Theoretical and Philosophical Psychology, 20*(2), 140–166.

Barden, N. (2004). Implication of the hypothalamic–pituitary–adrenal axis in the physiopathology of depression. *Journal of Psychiatry and Neuroscience, 29*(3), 185–193.

Barinaga, M. (2003). Buddhism and neuroscience. Studying a well trained mind. *Science, 302,* 5642–5644.

Barr, S. M. (2006). *Modern physics and ancient faith.* Notre Dame, IN: University of Notre Dame Press.

Bartlett, F. (1932). *Remembering* . Cambridge, MA: Cambridge University Press.

Barrett, L. F. (2009). The future of psychology: Connecting mind to brain. *Perspectives on Psychological Science, 4*(4), 326–339.

Barrett, L. F., Bliss-Moreau, E., Duncan, S. L., Rauch, S. L., & Wright, C. I. (2007). The amygdala and the experience of affect. *Social Cognitive and Affective Neuroscience, 2*(2), 73–83.

Barrett, L. F., Mesquita, B., Ochsner, K. N., & Gross, J. J. (2007). The experience of emotion. *Annual Review of Psychology, 58,* 373–403.

Bateson, G. (1972/2000). *Steps to an ecology of mind: Collected essays in the anthropology, psychiatry, evolution and epistemology.* Chicago, IL: University of Chicago Press.

Baulieu, E. E. (1996). Dehydroepiandrosterone (DHEA): Fountain of youth? *Journal of Clinical Endocrinology Metabolism, 81,* 3147–3151.

Baulieu, E. E., & Robel, P. (1998). Dehydroepiandrosterone (DHEA) and dehydroepiandrosterone sulfate (DHEAS) as neuroactive neurosteroids. *Proceedings of the National Academy of Science, 95,* 4089–4091.

Bechtel, W., & Stuffebeam, R. S. (2001). Epistemic issues in procuring evidence about the brain. In W. Bechtel, R. Stuffle beam, J. Mundale, & P. Mandik (Eds.), *Philosophy and the neurosciences: A reader* (pp. 55–81). Malden, MA: Blackwell.

Benchenane, K., Tiesinga, P. H., & Battaglia, F. P. (2011). Oscillations in the prefrontal cortex: A gateway to memory and attention. *Current Opinion in Neurobiology.* Advance online publication. doi:10.1016/j.conb.2011.01.004

Bentley, F., Vuilleumier, P., Thiel, C. M., Driver, J., & Dolan, R. J. (2003). Cholinergic enhancement modulates neural correlates of selective attention and emotional processing. *NeuroImage, 20*(1), 58–70.

Berdyaev, N. (1944). *Slavery and freedom.* New York, NY: Scribner's.

Berdyaev, N. (1951) *Dream and reality: An essay in autobiography.* New York, NY: Palgrave Macmillan.

Berger, P. L., & Luckmann, T. (1966). *The social construction of reality: A treatise in the sociology of knowledge.* New York, NY: Anchor Books.

Bernstein, R. J. (1983). *Beyond objectivism and relativism: Science, hermeneutics, and praxis.* Philadelphia, PA: University of Pennsylvania Press.

Biswal, B. B., Mennes, M., Zuo, X. N., Gohel, S., Kelly, C., Smith, S. M.,…Milham, M. P. (2010). Toward discovery science of human brain function. *Proceedings of the National Academy of Sciences of the United States of America, 107*(10), 4734–4739.

Bitbol, M. (2003). A cure for metaphysical illusions: Kant, quantum mechanics, and the Madhyamaka (pp. 325–361). In B. A. Wallace (Ed.), *Buddhism and science: Breaking new ground.* New York, NY: Columbia University Press.

Blakemore, S. J., & Frith, U. (2005). *The learning brain: Lessons for education.* New York, NY: Wiley-Blackwell.

Bradley, A. J., & Dinan, T. G. (2010). A systematic review of hypothalamic–pituitary–adrenal axis function in schizophrenia: Implications for mortality. *Journal of psychopharmacology Oxford England, 24*(11), 91–118.

Bredo, E. (1994). Reconstructing educational psychology: Situated cognition and Deweyian pragmatism. *Educational Psychologist, 29*(1), 23–35.

Brefczynski-Lewis, J. A., Lutz, A., Schaefer, H. S., Levinson, D. B., & Davidson, R. (2007). Neural correlates of attentional expertise in long-term meditation practitioners. *Proceedings of the National Academy of Sciences, 104*(27), 11483–11488.

Brennan, J. F. (1994). *History and systems of psychology* (4th ed.). Englewood Cliffs, NJ: Prentice Hall.

Breunig, M. (2005). Turning experiential education and critical pedagogy theory into praxis. *Journal of Experiential Education, 28*(2), 106–122.

Brockington, I. F., & Meakin, C. J. (1994). Clinical clues to the aetiology of puerperal psychosis. *Progressive Neuropsychopharmacology, Biology, and Psychiatry, 18,* 417–429.

Bruner, J. S. (1961). *A study of thinking.* New York, NY: John Wiley and Sons, Inc.

Bruner, J. S. (1968). *Toward a theory of instruction.* New York, NY: W.W. Norton & Co.

Bruner, J. S. (1973). *Beyond the information given: Studies in the psychology of knowing.* New York, NY: W.W. Norton & Co.

Bruner, J. S. (1993). *Acts of meaning.* Cambridge, MA: Harvard University Press.

Bruner, J. S. (1996). *The culture of education.* Cambridge, MA: Harvard University Press.

Bruner, J. S., Goodnow, J. J., & Austin, G. (1956). *A study of thinking.* New Brunswick, NJ: John Wiley and Sons Inc., Transaction Publishers.

Bugental, J. F. T. (1963). Humanistic psychology: A new breakthrough. *American Psychologist, 18,* 563–567.

Bühler, C. M. (1968). The general structure of the human life cycle. In C. Bühler & F. Massarik (Eds.), *The course of human life: A study of goals in the humanistic perspective* (pp. 12–26). New York, NY: Springer.

Burger, H. G. (2002). Androgen production in women. *Fertility and Sterility, 77*(4), S3-S5.

Burr, V. (2002). *The person in social psychology.* New York, NY: Taylor & Francis.

Burtis, C. A., Ashwook, E. R., & Bruns, D. E. (2011). *Tietz textbook of clinical chemistry and molecular diagnostics* (5th ed.). Philadelphia, PA: W. B. Saunders.

Buytendijk, F. J. J. (1967). Husserl's phenomenology and its significance for contemporary psychology. In N. Lawrence & D. O'Connor (Eds.), *Readings in existential-phenomenology.* Englewood Cliffs, NJ: Prentice Hall.

Buytendijk, F. J. J. (1974). *Prolegomena to an anthropological physiology.* Pittsburgh, PA: Duquesne University Press.

Buzsáki, G. (2006). *Rhythms of the brain.* New York, NY: Oxford University Press.

Cahn, B. R., Delorme, A., & Polish, J. (2010). Occipital gamma activation during Vipassana meditation. *Cognitive Process, 11,* 39–56.

Cahn, B. R., & Polich, J. (2006). Meditation states and traits: EEG, ERP, and neuroimaging studies. *Psychological Bulletin, 132*(2), 180–211.

Cai, W. (2012). Between the sense of self and the reality of self. *Phenomenology and the Cognitive Sciences, 11,* 113–118.

Caine, R. N., & Caine, G. (1991). *Making connections: Teaching and the human brain.* Menlo-Park, CA: Innovative Learning Publications.

Camic, P. J., Rhodes, J., & Yardley, L. (Eds.), *Qualitative research in psychology: Expanding perspectives in methodology and design* (pp. 243–273). Washington, D.C.: American Psychological Association.

Campbell, J. (1949). *The hero with a thousand faces.* New York, NY: Bollingen Foundation.

Canli, T. (2008). Towards a neurogenetic theory of neuroticism. *Annals of the New York Academy of Science, 1129*(1), 153–174.

Canli, T., Congdon, E., Constable, R. T., & Lesch, K. P. (2008). Additive effects of serotonin transporter and tryptophan hydrolase-2 gene variation in neural correlates of affective processing. *Biological Psychology, 79,* 118–125.

Canli, T., Zhao, Z., Brewster, J., Gabrieli, J. D., & Cahill, L. (2000). Event related activation in the human amygdala associates with later memory for individual emotional experience. *Journal of Neuroscience, 20,* RC99.

Cannon, W. B. (1931). Again the James-Lange and the thalamic theories of emotion. *The Psychological Review, 38*(4), 281–295.

Caputo, J. D. (1987). *Radical hermeneutics: Repetition, deconstruction and the hermeneutic project*. Bloomington, IN: Indiana University Press.

Caracciolo, M. (2012). Narrative, meaning, interpretation: An enactivist approach. *Phenomenology and the Cognitive Sciences, 11,* 367–384.

Carruthers, P., & Veillet, B. (2012). The case against cognitive phenomenology. In T. Bayne & M. Montague (Eds.), *Cognitive phenomenology* (pp. 35–55). New York, NY: Oxford University Press.

Chan, D., & Woolacott, M. (2007). Effects of level of meditation experience on attentional focus: Is the efficiency of executive or orientation networks improved? *The Journal of Alternative and Complementary Medicine, 13*(6), 651–657.

Changeux J-P., & Danchin, A. (1976). Selective stabilisation of developing synapses as a mechanism for the specification of neuronal networks. *Nature, 264,* 705–712.

Changeux, J-P., & Ricoeur, P. (2000). *What makes us think? A neuroscientist and a philosopher argue about ethics, human nature, and the brain* . Princeton, NJ: Princeton University Press.

Chen, T. L., Babiloni, C., Ferretti, A., Perrucci, M. G., Romani, G. L., Rissini, P. M.,… Del Gratta, C. (2010). Effects of somatosensory stimulation and attention on human somatosensory cortex: an fMRI study. *Neuroimage, 53*(1), 181–188.

Chiesa, A. (2010). Vipassana meditation: Systematic review of current evidence. Journal of *Alternative and Complementary Medicine, 16*(1), 37–46.

Cho, J., & Trent, A. (2006). Validity in qualitative research. *Qualitative Research, 6*(3), 319–340.

Choi, D. D., Furay, A. R., Evanson, N. K., Ostrander, M. M., Ulrich, Y. M., & Herman, J. P. (2007). Bed nucleus of the stria terminalis subregions differentially regulate hypothalamic-pituitary adrenal axis activity: Implications for the integration of limbic inputs. *The Journal of Neuroscience, 27*(8), 2025–2034.

Churchill, S. D. (2010). "Second person" perspectivity in observing and understanding emotional expression. In L. Embree, M. Barber, & T. J. Nenon (Eds.), *Phenomenology/Selected essays from North America, Part 2: Phenomenology beyond Philosophy* (Vol. 5). Bucharest/Paris: Zeta Book/Arghos-Diffusion.

Churchill, S. D., & Wertz, F. J. (2001). An introduction to phenomenological research in psychology: Historical, conceptual, and methodological foundations. In K. J. Schneider, J. F.T. Bugental, & J. F. Pierson (Eds.), *The handbook of humanistic psychology: Leading edges in theory, research and practice* (pp. 247–262).Thousand Oaks, CA: Sage Publications.

Churchland, P. S. (1986). *Neurophilosophy.* Cambridge, MA: The MIT Press.

Cilliers, P. (1998). *Complexity and postmodernism: Understanding complex systems.* New York, NY: Routledge.

Clancey, W. J. (2008). The scientific antecedents of situated cognition. In P. Robbins & M. Aydede (Eds.), *The Cambridge handbook of situated cognition* (pp. 11–34). Cambridge, MA: Cambridge University Press.

Clark, A. (1997). *Being there: Putting brain, body, and the world together again.* Cambridge MA: The MIT Press.

Clark, A. (2001). *Mindware: An introduction to the philosophy of cognitive science.* New York, NY: Oxford University Press.

Clegg, J. W. (2006). Phenomenology as foundational to the naturalized consciousness. *Culture and Psychology, 12*(3), 340–351.

Clegg, J. W. (2009). The problems of reductionism in the social sciences. In J. W. Clegg (Ed.), *The observation of human systems: Lessons from the history of anti-reductionistic empirical psychology* (pp. 1–8). New Brunswick, NJ: Transaction Publishers.

Collins, A., Eneroth, P., & Landgren, B. (1985). Psychoneuroendocrine stress responses and mood as related to the menstrual cycle. *Psychosomatic Medicine, 47*(6), 512–527.

Costall, A. (1995). Socializing affordances. *Theory & Psychology, 5,* 467–481.

Costall, A., & Still, A. W. (1991). Introduction: Cognitivism as an approach to cognition. In A. Still, & A. Costall (Eds.), *Against cognitivism: Alternative foundations for cognitive psychology* (pp. 1–6). Hertsfordshire, UK: Harvester Wheatsheaf.

Cowan, N. (2001). The magical number 4 in short-term memory: A reconsideration of mental storage capacity. *Behavioral Brain Science, 24,* 87–118.

Cox, R. F. A., & Smitsman, A. W. (2008). Towards an embodiment of goals. *Theory & Psychology, 18*(3), 317–340.

Craig, A. D. (2010). The sentient self . *Brain Structure and Function, 214*(5–6), 563–77.

Cranton, P. (1996). *Professional development as transformative learning: New perspectives for teachers of adults.* San Francisco, CA: Jossey-Bass.

Cubała, W. J., & Landowski, J. (2006). Serotoninergic system and limbic-hypothalamic-pituitary adrenal axis (LHPA axis) in depression. *Psychiatria Polska, 40*(3), 415–430.

Daniels, H. (2008). *Vygotsky and research.* New York, NY: Routledge.

Davanger, S., Ellingsen, O. Holen, A., & Higdahl, K. (2010). Meditation-specific prefrontal cortical activation during acem meditation: An fMRI study. *Perception & Motor Skills, 111*(1), 291–306.

David, N., Bewernick, B. H., Cohen, M. X., Newen, A., Lux, S., Fink, G. R.,…Vogeley, K. (2006). Neural representatives of self versus other: Visual-spatial perspective taking and agency in a virtual ball-tossing game. *Journal of Cognitive Neuroscience, 18*(6), 898–910.

Davidson, R. J., Goleman, D. J., & Schwartz, G. E. (1976). Attentional and affective concomitants of meditation: A cross-sectional study. *Journal of Abnormal Psychology, 85*(2), 235–238.

Davis, A. B., & Sumara, D. J. (2000). Curriculum forms: On the assumed shapes of knowing and knowledge. *Journal of Curriculum Studies, 32,* 821–845.

Davis, B., Sumara, D., & Luce-Kaplan, R. (2005). *Engaging minds. Learning and teaching in a complex world* . Mahwah, NJ: Lawrence Erlbaum.

Dehaene, S., & Naccache, L. (2001). Towards a cognitive neuroscience of consciousness: Basic evidence and a workspace framework. *Cognition, 79*(1–2), 1–37.

De Jaegher, H., & Di Paolo, E. A. (2007). Participatory sense-making: An enactive approach to social cognition. *Phenomenology and the Cognitive Sciences, 6*(4), 485–507.

Dennett, D. (1991). *Consciousness explained.* Boston, MA: Little, Brown and Co.

Depraz, N. (1999). The phenomenological reduction as praxis. *Journal of Consciousness Studies, 6*(2–3), 95–110.

De Preester, H. (2005). *Body image and body schema: Interdisciplinary perspectives on the body.* Philadelphia, PA: John Benjamins Publishing Company.

De Preester, H. (2007). The deep bodily origins of the subjective perspective: Models and their problems. *Consciousness and Cognition, 16,* 604–618.

DeRobertis, E. M. (2011). Existential-humanistic and dynamic systems approaches to child development in mutual encounter. *The Humanistic Psychologist, 39,* 3–23.

Derrida, J. (1973). Difference. In *Speech and phenomena and other essays on Husserl's theory of signs* (D. B. Allison & N. Garver, Trans.). Evanston, IL: Northwestern University Press.

Descartes, R. (1980). *Discourse on method and meditations on first philosophy.* Indianapolis, IN: Hackett Publishing Co. (Original work published 1637/1641)

Descombes, V. (2001). *The mind's provisions: A critique of cognitivism* (S. A. Schwartz, Trans.). Princeton, NJ: Princeton University Press.

Dewey, J. (1938). *Experience and education.* New York, NY: Simon and Schuster.

Dewey, J., & Bentley, A. F. (1949). *Knowing and the known.* Boston, MA: Beacon Press.

Dillon, M. C. 1997. *Merleau-Ponty's ontology* (2nd ed.). *Northwestern University studies in phenomenology and existential philosophy*. Evanston, Ill: Northwestern University Press.

Di Paolo, E. (2009). Editorial: The social and enactive mind. *Phenomenology and the Cognitive Sciences, 8*, 409–415.

Di Paolo, E. A., Rohde, M., & Jaegher, H. (2007). Horizons for the enactive mind: Values, social interaction, and play. In J. Stewart, O. Gapenne, & E. A. Paolo (Eds.), *Enaction: Towards a new paradigm for cognitive science*. Cambridge, MA: The MIT Press.

Dirkx, J. M. (2002). The power of feelings: Emotion, imagination, and the construction of meaning in adult learning. *New Directions for Adult and Continuing Education, 89*, 63–72.

Doucet, G., Naveau, M., Petit, L., Delcroix, N., Zago, L., Crivello, F.,…Joliot, M. (2011). Brain activity at rest: A multi-scale hierarchical functional organization. *Journal of Neurophysiology*. Advance online publication. doi:10.1152/jn.00895.2010

Dreyfus, H. L. (1972). *What computers can't do: A critique of artificial reason*. New York, NY: Harper & Row.

Dreyfus, H. L. (1982). Introduction. In H. Dreyfus (Ed.), *Husserl, intentionality and cognitive science*. Cambridge, MA: The MIT Press.

Dreyfus, H. (1992). *What computers can't do: A critique of artificial intelligence*. New York, NY: Harper and Row. (Original work published 1979)

Ellenberger, H. (1970). *The discovery of the unconscious*. New York, NY: Basic Books.

Ellis, R. D. (1999). Integrating neuroscience and phenomenology in the study of consciousness *Journal of Phenomenology, 30*(1), 18–47.

Ellis, R. D. (2006). Phenomenology-friendly neuroscience: The return to Merleau-Ponty as psychologist. *Human Studies, 29*(1), 33–55.

Erfani, A., & Erfanian, A. (2004). The effects of mental practice and concentration skills on EEG brain dynamics during motor imagery using independent component analysis. *Conference Proceedings of the IEEE Engineering in Medicine and Biology Society, 1*, 239–42.

Etkin, A., Klemenhagen, K. C., Dudman, J. T., Rogan, M. T., Hen, R., & Kandel, E. R.,…Hirsch, J. (2004). Individual differences in trait anxiety predict the response of the basolateral amygdala to unconsciously processed fearful faces. *Neuron 44*(6), 1043–1055.

Fadiga, L., Fogassi, L., Pavesi, G., & Rizzolatti, G. (1995). Motor facilitation during action observation: A magnetic stimulation study. *Journal of Neurophysiology, 73*, 2608–2611.

Fan, J., Byrne, J., Worden, M. S., Guise, K. G., McCandliss, B. D., Fossella, J., & Posner, M. I. (2007). The relation of brain oscillations to attentional networks. *The Journal of Neuroscience, 27*(23), 6197–6206.

Fasching, W. (2008). Consciousness, self-consciousness, and meditation. *Phenomenology and the Cognitive Sciences, 7*, 463–483.

Feinstein, D., Krippner, S., & Granger, D. (1988). Mythmaking and human development. *Journal of Humanistic Psychology, 28*, 23–50.

Fell, J., Axmacher, N., & Haupt, S. (2010). From alpha to gamma: Electrophysiological correlates of meditation-related states of consciousness. *Medical Hypothesis, 75*(2), 218–224.

Fendler, L. (1998). What is impossible to think? A genealogy of the educated subject. In T. S. Popkewitz, & M. Brennan (Eds.), *Foucault's challenge: Discourse, knowledge, and power in education* (pp. 39–62). New York, NY: Teachers College Press.

Fenwick, T. (2000). Expanding conceptions of experiential learning: A review of the five contemporary perspectives on cognition. *Adult Education Quarterly, 50*(4), 243–273.

Finlay, L. (2008). A dance between the reduction and reflexivity: Explicating the "Phenomenological Psychological Attitude." *Journal of Phenomenological Psychology, 39*, 1–32.

Finlay, L. (2009). Debating phenomenological research methods. *Phenomenology & Practice, 3*(1), 6–25.

Fisher, A. M. (1997). Modern manifestations of materialism: A legacy of the enlightenment discourse. *Journal of Theoretical and Philosophical Psychology, 12*(1), 45–55.

Fodor, J. (1981). *Representations: Philosophical essays on the foundations of cognitive science*. Cambridge, MA: The MIT Press.

Forgas, J. P. (2002). Affective influences on attitudes and judgments (pp. 596–618). In R. J. Davison, K. R. Scherer, & H. H. Goldsmith (Eds.), *Handbook of affective sciences*. New York, NY: Oxford University Press.

Forman, R. K. C. (1990). Introduction: Mysticism, constructivism and forgetting. In R. K. C. Forman (Ed.), *The problem of pure consciousness* (pp. 3–49). New York, NY: Oxford University Press.

Foucault, M. (1980). Two lectures in power and knowledge. In C. Gordon (Ed.), *Power/knowledge: Selected interviews and other writings* (pp. 78–108). New York, NY: Harvester Press.

Frankl, V. E. (1959). *From death-camp to existentialism: A psychiatrist's path to a new therapy.* Boston, MA: Beacon Press.

Frankl, V. E. (1966). Self-transcendence as a human phenomenon. *Journal of Humanistic Psychology, 6,* 97–106.

Freeman, W. J. (1999). *How brains make up their minds.* London, UK: Weidenfeld & Nocolson.

Freeman, W. J. (2000). Emotion is essential to all intentional behaviors. In M. D. Lewis & I. Granic (Eds.), *Emotion, development and self-organization: Dynamic systems approaches to emotional development* (pp. 209–235). New York, NY: Cambridge University Press.

Freeman, W. J. (2009). Vortices in brain activity: Their mechanism and significance for perception. *Neural Networks, 22,* 491–501.

Freire, P. (1971). *Pedagogy of the oppressed.* New York, NY: Seabury Press.

Fridland, E. (2011). The case for proprioception. *Phenomenology and the Cognitive Sciences, 10,* 521–540.

Frie, R. (Ed.). (2003). *Understanding experience: Psychotherapy and postmodernism.* New York: NY: Routledge.

Froese, T., & Fuchs, T. (2012). The extended body: A case study in the neurophenomenology of social interaction. *Phenomenology and the Cognitive Sciences, 11,* 205–235.

Fujiki, M., Akiyoshi, J., Yoshida, T., Tabata, M., Kabsawa, H., Kobayashi, H., & Shimomura, T. (2008). Functional brain mapping during recitation of Buddhist scriptures and repetition of the Namu Amida Butsu; A study in experienced Japanese monks. *Turkish Neurosurgery, 18*(2), 134–141.

Gadamer, H. (1976). *Philosophical hermeneutics* (D. Lingis, Trans., Ed.). Berkeley, CA: University of California Press.

Gallagher, S. (1997). Mutual enlightenment: Recent phenomenology in cognitive sciences. *Journal of Consciousness Studies, 4*(3), 195–214.

Gallagher, S. (1998). *The inordinance of time*. Evanston, Ill: Northwestern University Press.

Gallagher, S. (2003). Phenomenology and experimental design. *Journal of Consciousness Studies, 10*(9–10), 85–99.

Gallagher, S. (2004). Hermeneutics and the cognitive sciences. *Journal of Consciousness Studies, 11*(10–11), 162–174.

Gallagher, S. (2005). *How the body shapes the mind.* London, UK; New York, NY: Oxford University Press.

Gallagher, S. (2007). Neurophilosophy and neurophenomenology. In L. Embree & T. Nelson (Eds.), *Phenomenology* (Vol. 5, pp. 293–316). Bucharest, Romania: Zeta Press.

Gallagher, S. (2009). Philosophical antecedents of situated cognition. In P. Robbins, & M. Aydede (Eds.), *The Cambridge handbook of situated cognition.* Cambridge, MA: Cambridge University Press.

Gallagher, S., & Sorensen, J. B. (2006). Experimenting with phenomenology. *Consciousness and Cognition, 15,* 119–134.

Gallagher, S., & Zahavi, D. (2008). *The phenomenological mind: An introduction to philosophy of mind and cognitive science.* New York/Abingdon/Oxon: Routledge.

Gallese, V. (1998). Mirror neurons: From grasping to language. Paper presented at Tucson III Conference: Towards a Science of Consciousness.

Gallese, V. (2001). The "shared manifold hypothesis:" From mirror neurons to empathy. *Journal of Consciousness Studies, 8*(5–7), 33–50.

Gallese, V., Fadiga, L., & Rizzolatti, G. (1996). Action recognition in premotor cortex. *Brain, 119,* 593–609.

Garcia, Varela, F. J. (1970). *Insect Retinas: Visual processing in the compound eye.* (Unpublished doctoral dissertation). Harvard University, Cambridge, MA (HU 90.9880.15).

Gardner, H. (1983). *Frames of mind and the theory of multiple intelligences.* New York, NY: Basic Books.

Gardner, H. (1991). *The unschooled mind: How children think and how schools should teach.* New York, NY: Basic Books.

Gardner, H. E. (1987). *The mind's new science: A history of the cognitive revolution.* New York, NY: Basic Books.

Gendlin, E. T. (1978a). *Focusing.* New York, NY: Bantam Books.

Gendlin, E. T. (1978b). *Befindlichkeit:* Heidegger and the philosophy of psychology. *Review of Existential Psychology and Psychiatry, 16*(1–3), 43–71.

Gergen, K. (1995). Social construction and the educational process. In L. P. Steffe, & J. Gale (Eds.), *Constructivism in education.* Hillsdale, NJ: Lawrence Erlbaum Associates.

Gibbs, R. W., & Van Orden, G. C. (2003). Are emotional expressions intentional? A self organizational approach. *Consciousness and Emotion, 4,* 1–16.

Gibson, J. J. (1979). *The ecological approach to visual perception.* Boston, MA: Houghton Mifflin.

Giorgi, A. (1970). *Psychology as a human science: A phenomenologically based approach.* New York, NY: Harper & Row.

Giorgi, A. (1975). An application of phenomenological method in psychology. In A. Giorgi, C. Fischer, & E. Murray (Eds.), *Duquesne studies in phenomenological psychology: Volume II* (pp. 72–79). Pittsburgh, PA: Duquesne University Press.

Giorgi, A. (1997). The theory, practice, and evaluation of the phenomenological method as a qualitative research procedure. *Journal of Phenomenological Psychology, 28*(2), 235–260.

Giorgi, A. (2009a). *The descriptive phenomenological method in psychology.* Pittsburgh, PA: Duquesne University Press.

Giorgi, A. (2009b, May, 1). [Review of the book *The phenomenological mind: An introduction to philosophy of mind and cognitive science*, by S. Gallagher and D. Zahavi]. *Journal of Phenomenological Psychology, 40*(1), 107–125.

Giorgi, A. (2010). Phenomenological psychology: A brief history and its challenges. *Journal of Phenomenological Psychology, 41,* 145–179.

Giroux, H. A. (2001). *Theory and resistance in education: Towards a pedagogy for the opposition.* Westport, CT: Bergin & Garvey.

Globus, G. G. (1995). *The postmodern brain: Advances in consciousness research.* Philadelphia, PA: John Benjamins Publishing Company.

Goldstein, J. M. (2006). Sex, hormones and affective arousal circuitry dysfunction in schizophrenia. *Hormones and Behavior, 50,* 612–622.

Goldstein, J. M., Jerram, M., Poldrack, R., Ahern, T., Kennedy, D. N., & Seidman, L. J.,...Markis, N. (2005). Hormonal cycle modulates arousal circuitry in women using functional magnetic resonance imaging. *The Journal of Neuroscience, 25*(40), 9309–9316.

Goleman, D. (1988). *The meditative mind: The varieties of meditative experience.* New York, NY: Penguin, Putman.

Goleman, D. (1995). *Emotional intelligence: Why it can matter more than I.Q.* New York, NY: Bantam Books.

Good, J. M. M. (2007). The affordances for social psychology of the ecological approach to social knowing. *Theory & Psychology, 17*(2), 265–295.

Gordon, D. (1988). Tenacious assumptions in western medicine. In M. Lock & D. R. Gordon (Eds.), *Biomedicine examined* (pp. 19–56). New York, NY: Kluwer Academic Publishers.

Gordon, S. (2001). *A theoretical investigation of the relationship between transpersonal self concept and estrogen dominance* . Unpublished master's thesis. Saybrook Graduate School and Research Center, San Francisco, CA.

Gordon, S. (2007). Psycho-neuro-intracrinology: The embodiment of myth, intentionality, and the spiritual sense of well-being in women at midlife. *Dissertation Abstracts International, B 68/09*. (UMI No. 3285857)

Gordon, S. (2009a, September). The mind and the brain [Review of the book *The Human Amygdala*]. *PsycCritiques – Contemporary Psychology: APA Review of Books, 54* (No. 39), Article 9. PsycCRITIQUES database.

Gordon, S. (2009b). Radical empiricism and the neurophenomenology of time. Unpublished working paper.

Gordon, S. (2012, February). Existential time and the meaning of human development. *The Humanistic Psychologist, 40* (1), 79–86.

Gordon, S. (2013, April). Internal time is autopoietic. [Review of the book *Internal Time: Chronotypes, Social Jet Lag, and Why You're So Tired]. PsycCritiques – Contemporary Psychology: APA Review of Book,* 58 (No.15), Article 4. PsycCRITIQUES database.

Gordon, S. (in press a). A neurophenomenological psychology of self. In R. Brook, C. T. Fischer, & L. Laubscher, (Eds.), *Invitation to psychology as a human science*. Pittsburgh, PA: Duquesne University Press.

Gordon, S. (in press b). Psycho-Neuro-Intracrinology: The mind-body continuum. In P. Snider, J. Zeff, J. Sensenig, & J. E. Pizzorno (Eds.), *The healing power of nature: The foundations of naturopathic medicine and the ecology of healing: Primary care for the twenty first century*. Philadelphia, PA: Elsevier.

Grafton, S. T., Arbib, M. A., Fadiga, L., & Rizzolatti, G. (1996). Localization of grasp representations in humans by PET 2: Observation compared with imagination. *Experimental Brain Research, 112,* 103–111.

Green, E., Green, A. M., & Walters, E. D. (1970).Voluntary control of internal states: Psychological and physiological. *Journal of Transpersonal Psychology, 2,* 1–26.

Green, R., & Turner, G. (2010). Growing evidence for the influence of meditation on brain and behavior. *Neuropsychological Rehabilitation, 20*(2), 306–311.

Greene, M (1995). *Releasing the imagination: Essays on education, the arts, and social change.* San Francisco, CA: Jossey-Bass Publishers.

Greeno, J. G. (1998). The situativity of knowing, learning and research. *American Psychologist, 53*(1), 5–26.

Gross, C. G. (2000). Neurogenesis in the adult brain: Death of a dogma. *Nature Reviews Neuroscience, 1,* 67–73.

Guenther, H. (1992). *Meditation differently.* Delhi, India: Motilal Banarsidas.

Gurwitsch, A. (1966). The place of psychology in the system of sciences. In A. Gurwitsch (Ed.), *Studies in phenomenology and psychology.* Evanston, IL: Northwestern University Press.

Hardy, L. (Trans.) (1990). *Edmund Husserl: The idea of phenomenology: A translation of Die Idee der Phänomenologie Husserliana II.* Dordrecht, Boston, London: Kluwer Academic Publishers.

Handley, K., Sturdy, A., Fincham, R., & Clark, T. (2006). Within and beyond communities of practice: Making sense of learning through participation, identity and practice. *Journal of Management Studies, 43*(3), 641–653.

Hare, T. A., Tottenham, N., Davidson, M. C., Glover, G. H., & Casey, B. J. (2005). Contributions of amygdale and striatal activity in emotion regulation. *Biological Psychiatry, 57*(6), 624–632.

Healey, J. M. (2004). *Your child's growing mind: Brain development and learning from birth to adolescence.* New York, NY: Broadway Books.

Hebb, D. O. (1949). *The organization of behaviour.* New York, NY: John Wiley & Sons.

Heidegger, M. (1927/1962). *Sein und zeit.* (J. Macquarrie and E. Robinson, Trans.). [Being and time]. New York, NY: Harper and Row.

Heidegger, M. (1962). *Being and time.* New York, NY: HarperCollins.

Heidegger, M. (1996). *Being and time* (J. Stambaugh, Trans.). New York, NY: State University of New York Press. (Original work published 1926)

Herman, J. P., Figueiredo, H., Mueller. K., Ulrich-Lai, Y., Ostrander, M. M., & Choi, D. C.,… Cullinan, W. E. (2003). Central mechanisms of stress integration: Hierarchical circuitry controlling hypothalamo-pituitary-adrenocortical responsiveness. *Frontiers in Neuroendocrinology, 24,*151–180.

Herman, J. P., Ostrander, M. M., Mueller, N. K., & Figueiredo, H. (2005). Limbic system mechanisms of stress regulation: hypothalamo-pituitary-adrenocortical axis. *Progress in Neuropsychopharmacology and Biological Psychiatry, 29*(8), 1201–1213.

Hill, L. H. (2001). The brain and consciousness: Sources of information for understanding adult learning. *New Directions for Adult and Continuing Education, 89,* 73–82.

Holland, P. C., & Gallagher, M. (1999). Amygdala circuitry in attentional and representational processes. *Trends in Cognitive Sciences, 3* (2), 65–73.

Holzbauer, M. (1976). Physiological aspects of steroids with anesthetic properties. *Medical Biology, 54,* 227–242.

Holzbauer, M., Birmingham, M. K., DeNicola, A. F., & Oliver, J. T. (1985). In vivo secretion of 3 alpha-hydroxy-5 alpha-pregnan-20-one, a potent unaesthetic steroid, by the adrenal gland of the rat. *Journal of Steroid Biochemistry, 22*(1), 97–102.

hooks, B. (1994). *Teaching to transgress: Education as the practice of freedom.* New York, NY: Routledge.

Hoshmand, L.T., & Martin, J. (1994). Naturalizing the epistemology of psychological research. *Journal of Theoretical and Philosophical Psychology, 14*(2), 171–189.

Huang, H. Y., & Lo, P. C. (2009). EEG dynamics of experienced Zen meditation practitioners probed by complexity index and spectral measure. *Journal of Medical Engineering and Technology, 33*(4), 314–321.

Husserl, E. (1900–1901). *Logische Untersuchungen. Zweiter Teil: Untersuchungen zur Phänomenologie und Theorie der Erkenntnis.* [*Logical investigations* . 2 vols.: *Studies in phenomenology*; *The theory of knowledge* .] Halle a.S.: Max Niemeyer.

Husserl, E. (1950/1966). *Phänomenologie des inneren zeitbewu b tseins.* (J. S. Churchill, Trans.). [The Phenomenology of Internal Time Consciousness]. Bloomington, IN: University of Indiana Press. (Original work published 1893–1917)

Husserl, E. (1962). *Ideas: General introduction to pure phenomenology* (W. R. B. Gibson, Trans.). New York, NY: Collier Books. (Original work published 1913)

Husserl, E. (1973). *Logical investigations* (J. N. Finlay, Trans.). New York, NY: Routledge. (Original work published 1900)

Husserl, E. (1977). *Phenomenological psychology* (J. Scanlon, Trans.) The Hague, Netherlands: Martinus Nijhoff. (Original work published 1925)

Husserl, E. (1980). Ideas pertaining to a pure phenomenology and to a phenomenological philosophy. *Collected works Edmund Husserl.* The Hague, Boston, Hingham, MA, USA: M. Nijhoff.

Husserl, E. (1982). *Ideas pertaining to a pure phenomenology and to a phenomenological philosophy – First book: General introduction to a pure phenomenology* (F. Kersten, Trans.). The Hague: Nijhoff. (Original work published 1913)

Husserl, E. (1991). *Collected Works. Vol. 4: On the phenomenology of the consciousness of internal time* (1893–1917) (J. Brough, Trans.). Dordrecht: Kluwer Academic.

Husserl, E. (1998). *Ideas pertaining to a pure phenomenology and to a phenomenological philosophy* (2nd ed.) (Rojcewicz, R. & Schuwer, A., Trans.). Dordrecht, The Netherlands: Kluwer Academic Publishers. (Original work published 1913)

Hutchins, E. (1995). *Cognition in the wild.* Cambridge, MA: The MIT Press.

Hwang, R-J., Wu, C-H., Chen, L-F., Yeh, T-C., & Hsieh, J-C. (2009). Female menstrual phases modulate human prefrontal asymmetry: A magnoencephalographic study. *Hormones and Behavior, 55,* 203–209.

Ihde, D. (1986). *Experimental phenomenology: An introduction.* New York, NY: State University of New York Press.

Ihde, D. (1993). *Postphenomenology: Essays in the postmodern context.* Evanston, IL: Northwestern University.

Illeris, A. (2008). Transformative learning in the perspective of a comprehensive learning theory. *Journal of Transformative education, 2*(2), 78–89.
Immordino-Yang, M. H., & Fischer, K.W. (2011). Neuroscience bases of learning. In V. G. Aukrust (Ed.), *Learning and cognition in education* (pp. 9–15). London, UK: Elsevier Academic Press.
Ingold, T. (2001). From transmission of representations to the education of attention. In H. Whitehouse (Ed.), *The debated mind: Evolutionary psychology versus ethnography.* New York, NY: Berg, Oxford International Press.
Jain, M. R., Burade, V. S., & Subhedar, N. (1995). Neurosteroids: Novel regulators of neural functions. *Indian Journal of Pharmacology, 27,* 212–221.
James, W. (1865). Review of Alfred R. Wallace's The origin of human races and the antiquity of man deduced from the theory of 'natural selection.' *North American Review, 100,* 290–298.
James, W. (1868). Review (unsigned) of Charles Darwin's The variation of animals and plants under domestication. *Atlantic Monthly, 22,* 122–124.
James, W. (1874). Notice (unsigned) of J. B. Petigrew's Physiology of the circulation in plants, in lower animals, and in man. *Nation, 19,* 63.
James, W. (1875). Notice (unsigned) of F. Papillon's Nature and life. *Nation, 20,* 429.
James, W. (1876). Note (unsigned) on the *Revue Philosophique. Nation, 22,* 147.
James, W. (1877). Review (unsigned) on the Mind and the Brain (D. Ferrier, H. Maudsley, & J. Luys). *Nation, 2J h,* 355–356.
James, W. (1879a). Are we automata? *Mind, 4,* 1–22.
James, W. (1879b). The sentiment of rationality. *Mind, 4,* 317–346.
James, W. (1884). What is an emotion? *Mind, 9*(34), 188–205.
James, W. (1885). The function of cognition. *Mind, 10,* 27–44.
James, W. (1890). *The principles of psychology* (2 Vols.). New York, NY: Henry Holt.
James, W. (1892). *Psychology: Briefer course.* New York, NY: Henry Holt.
James, W. (1895). The tigers in India. *Psychological Review, 2,* 105.
James, W. (1897). *The will to believe and other essays in popular philosophy.* New York, NY: Longmans, Green & Co.
James, W. (1898, September). Philosophical conceptualizations and practical results. *University of California Chronicle, 1* (4), 24.
James, W. (1902). *The varieties of religious experience: A study in human nature*. New York, NY: Longmans, Green and Co.
James, W. (1904a). A world of pure experience. *Journal of Philosophy, Psychology and Scientific Methods, 1*(21), 533 – 543; 561–570.
James, W. (1904b). Does "consciousness" exist? *Journal of Philosophy, Psychology and Scientific Methods, 1*(18), 477–491.
James, W. (1907). *Pragmatism.* New York, NY: Longmans, Green, and Co.
James, W. (1912). *Essays in radical empiricism* (pp. 92–122). New York, NY: Longmans, Green and Co.
James, W. (1961). *Psychology briefer course.* New York, NY: Henry Holt. (Original work published 1892)
James, W. (1988). The 1878 Lowell Lectures. In F. Burkhardt, F. Bowers, & I. K. Skrupskeis (Eds.), *Manuscript lectures. The works of William James.* Cambridge, MA: Harvard University Press.
James, W. (2007). *Principles of psychology* (Vol. 1). New York, NY: Cosimo. (Original work published 1890)
Jang, J. H., Jung, W. H., Kang, D. H., Byun, M. S., Kwon, S. J., Choi, C. H., & Kwon, J. S. (2011). Increased default mode network connectivity associated with meditation. *Neuroscience Letters, 487* (3), 358–62.
Jankord, R., & Herman, J. P. (2008). Limbic regulation of hypothalamo-pituitary-adrenocortical function during acute and chronic stress. *Annals of the New York Academy of Science, 1148,* 64–73.
Jarvis, P. (2005). *Lifelong learning and the learning society complete trilogy set: Towards a comprehensive theory of human learning* (Vol. 1). New York, NY: Routledge.

Jarvis, P. (2006). *Towards a comprehensive theory of human learning.* New York, NY: Routledge.

Jensen, E. (1998). *Teaching with the brain in mind.* Alexandria, VA: Association for Supervision and Curriculum Development.

Jensen, E. (2008). *Brain-based learning: The new paradigm of teaching.* Thousand Oaks, CA: Corwin Press.

Jha, A. P., Krompinger, J., & Baime, M. J. (2007) Mindfulness training modifies subsystems of attention. *Cognitive, Affective and Behavioral Neuroscience, 7*(2), 109–119.

Jordi, R. (2011). Reframing the concept of reflection: Consciousness, experiential learning, and reflective learning practices. *Adult Education Quarterly, 61*(2), 181–197.

Josipovic, Z. (2010). Duality and nonduality in meditation research. *Consciousness and Cognition, 19*(4), 1119–1121.

Juarrero, A. (1999). *Dynamics in action: Intentional behavior as a complex system.* Cambridge, MA: The MIT Press.

Jung, C. G. (1969). *Psychology and religion: West and East* (2nd ed.) (R. F. C. Hull, Trans.). Princeton, NJ: Princeton University Press.

Jung, C. G. (1971). *On the nature of the psyche: Vol. 8, Collected works* (R. F. C. Hull, Trans.). Princeton, NJ: Princeton University Press. (Original work published 1946)

Jung. C. G. (2001). *Psychology of the unconscious: A study of the transformations and symbolisms of the libido: Collected works, Supplementary Vol. B.* (B. Hinkle, Trans.). Princeton, NJ: Princeton University Press. (Original work published 1911)

Kant, I. (2003). Transcendental doctrine of elements. In N. K. Smith (Trans.), *Critique of pure reason* (pp. 65–101). New York, NY: Palgrave Macmillan. (Original work published 1781)

Karmiloff-Smith, A. (1997). *Beyond modularity: A developmental perspective on cognitive science.* Cambridge, MA: The MIT Press.

Kincheloe, J. L. (1991). *Teachers as researchers: Qualitative inquiry as a path to empowerment.* London, UK; New York, NY: The Falmer Press.

Kincheloe, J. L. (1999). *Rethinking intelligence: Confronting psychological assumptions about teaching and learning.* London, UK: Routledge.

Kincheloe, J. L. (2005). *Critical constructivism.* New York, NY: P. Lang.

Kincheloe, J. L. (2008). *Critical pedagogy.* New York, NY: P. Lang.

Kinsbourne, M. (2006). From unilateral neglect to the brain basis of consciousness. *Cortex, 42,* 869–874.

Koch, S. (Ed). (1958–1964). *Psychology: The study of a science* (7 vols.). New York, NY: McGraw Hill.

Koestenbaum, V. (1997). *The phenomenological sense of John Dewey.* Atlantic Highlands, NJ: Humanities Press.

Köhler, W. (1938). *The mentality of apes.* New York, NY: Liveright.

Köhler, W. (1947). *Gestalt psychology.* New York, NY: Mentes.

Kosslyn, S. M. (1995). Mental imagery. In S. M. Kosslyn & D. N. Osherson (Eds.), *Visual cognition: An invitation to cognitive science* (Vol. 2) (pp. 267–296). Cambridge, MA: The MIT Press.

Kosslyn, S. M., Seger, C., Pani, J. R., & Hillger, L. A. (1990). When is imagery used in everyday life? A diary study. *Journal of Mental Imagery, 14*(3&4), 131–152.

Kozhevnikov, M., Louchakova, O., Josipovic, Z., & Motes, M. A. (2009). The enhancement of visuospatial processing efficiency through Buddhist deity meditation. *Psychological Science, 20*(5), 645–653.

Kravitz, D. J., Saleem, K. S., Baker, C. I., & Mishkin, M. (2011). A new neural framework for visuospatial processing. *Nature Reviews Neuroscience, 4,* 217–230.

Kuhn-Velten, W. N. (2000). Intracrinology and the local enzymatic control of hormone distribution and metabolism: Dehydroepiandrosterone does not just act as a prohormone for androgens and estrogens. *European Journal of Clinical Investigation, 30*(3), 34–38.

Kungurtsev, I., & Louchakova, O. (1994). The unknown Russian mysticism: Pagan sorcery, Christian Yoga, and other esoteric practices in the former Soviet Union. *Gnosis, 32,* 20–27.

Labrie, F. (1991). Intracrinology. *Molecular Cell Endocrinology, 78* (3), C113-C118.
Labrie, F. (2003). Extragonadal synthesis of sex steroids: Intracrinology. *Annals of Endocrinology* (Paris), *64*(2), 95–107.
Labrie, F. (2004). Adrenal androgens and intracrinology. *Seminars in Reproductive Medicine, 22*(4), 299–309.
Labrie, F., Bélanger, A., Cusan, L., & Candas, B. (1997). Physiological changes in dehydroepiandrosterone are not reflected by serum levels of active androgens and estrogens but of their metabolites: Intracrinology. *Journal of Clinical Endocrinology Metabolism, 82*(8), 2403–2409.
Labrie, F., Bélanger, A., Luu-The, V., Labrie, C., Simard, J., Cusan, L.,...Candas, B. (1998). DHEA and the intracrine formation of androgens and estrogens in peripheral target tissues: Its role during aging. *Steroids, 63*(5–6), 322–328.
Labrie, F., Bélanger, A., Simard, J., Luu-The, V., & Labrie, C. (1995). Intracrinology: Autonomy and freedom of peripheral tissues. *Annals of Endocrinology (Paris), 56*(1), 23–29.
Labrie, F., Luu-The, V., Bélanger, A., Lin, S. X., Simard, J., Pelletier, G.,...Labrie, C. (2005). Is dehydroepiandrosterone a hormone? *Journal of Endocrinology, 187*(2), 169–96.
Labrie, F., Luu-The, V., Labrie, C., & Simard, J. (2001). DHEA and its transformation into androgens and estrogens in peripheral target tissues: Intracrinology. *Frontiers in Neuroendocrinology, 22*(3), 185–212.
Lambie, J. A., & Marcel, A. J. (2002). Consciousness and the varieties of emotion experience: A theoretical framework. *Psychological Review, 109*(2), 219–259.
Langfur, S. (2012). The You-I event: On the genesis of self awareness. *Phenomenology and The Cognitive Sciences* , retrieved October 9, 2012 from http://link.springer.com/journal/11097. doi: 10.1007/s11097-012-9282-y
Larkin, M., Eatough, V., & Osborn, M. (2011). Interpretive phenomenological analysis and embodied, active, situated cognition. *Theory & Psychology, 21*(3), 318–337.
Lather, P. (1986). Research as praxis. *Harvard Educational Review, 56,* 257–277.
Lather, P. (1991). *Getting smart: Feminist research and pedagogy with/in the postmodern.* New York, NY: Routledge.
Laughlin, C., McManus, J., & d'Aquili, E. (1990). *Brain, symbol and experience: Toward a neurophenomenology of consciousness.* New York, NY: Columbia University Press.
Lave, J. (1988). *Cognition in practice.* Cambridge, MA: Cambridge University Press.
Lave, J., & Wenger, E. (1991). *Situated learning: Legitimate peripheral participation.* New York, NY: Cambridge University Press.
Lazar, S. W., Kerr, C. E., Wasserman, R. H., Gray, J. R., Greve, D. N., Treadway, M. T.,... Fischl, B. (2005). Meditation experience is associated with increased cortical thickness. *Neuroreport, 16*(17), 1893–1897.
Lazarus, R. S. (1994). *Emotion and adaptation.* New York, NY: Oxford University Press.
LeDoux, J. E. (1996). *The emotional brain: The mysterious underpinnings of emotional life.* New York, NY: Simon & Schuster.
LeDoux, J. E. (2000). Emotion circuits in the brain. *Annual Review of Neuroscience, 23*(1), 155–184.
LeDoux, J. (2002). *Synaptic self: How our brains become who we are.* New York, NY: Penguin Books.
Legrand, D. (2007). Pre-reflective self-as-subject from experiential and empirical perspectives. *Consciousness and Cognition, 16,* 583–599.
Leisman, G., & Koch, P. (2009). Networks of conscious experience: Computational neuroscience in understanding life, death, and consciousness. *Reviews in the Neurosciences, 20,* 151–176.
Lemke, J. L. (1997). Cognition, context, and learning: A social semiotic perspective. In D. Kirshner & J. Whitson (Eds.), *Situated cognition: Social, semiotic and psychological perspectives* (pp. 37–57). Mahwah, NJ: Lawrence Erlbaum Associates.
Lewis, M. D., & Todd, R. M. (2005). Getting emotional: A neural perspective on emotion, intention, and consciousness. *Journal of Consciousness Studies, 12*(8–10), 213–238.

Lindberg, D. A. (2005). Integrative review of research related to meditation, spirituality, and the elderly. *Geriatric Nursing, 26*(6), 372–377.

L'Institut des Systèmes Complexes Paris Île-de-France (ISC-PIF). (Jun, 2009). Preamble. *French Complex Systems Roadmap, 2008 – 2009* (hal-00392486, version 1–8, pp. 5–8). Paris, France: Complex Systems Institute.

Lo, P. C., Huang, M. L., & Chang, K. M. (2003). EEG alpha blocking correlated with perception of inner light during Zen meditation. *The American Journal of Chinese Medicine, 31*(4), 629–642.

Longchenpa. (1975). *Kindly bent to ease us. Part one: Mind* (H. V. Guenther, Trans.) Berkeley, CA: Dharma Publishing.

Lou, H. C., Kjaer, T. W., Friberg, L., Wildschiodtz, G., Holm, S., & Nowak, M. (1999). A 15OH2O PET study of meditation and the resting state of normal consciousness. *Human Brain Mapping, 7,* 98–105.

Louchakova, O. (2005). Ontopoiesis and union in the prayer of the heart: Contributions to psychotherapy and learning. In A. T. Tymieniecka (Ed.), *Analecta Husserliana: The yearbook of phenomenological research (Vol. 91). Logos of phenomenology and phenomenology of the logos. Book four: The logos of scientific interrogation. Participating in nature –life- sharing in life* (pp. 289–311). Dordrecht, UK: Kluwer.

Louchakova, O. (2007a). Ontopoiesis and spiritual emergence: Bridging Tymieniecka's phenomenology of life and transpersonal psychology. In A.-T. Tymieniecka (Ed.), *Analecta Husserliana: Phenomenology of life- from the animal soul to the human mind* (Vol. 94) (pp. 43–68). Dordrecht, UK: Kluwer.

Louchakova, O. (2007b). Prayer of the Heart, ego-transcendence and adult development. *Existential Psychoanalysis, 18*(2), 261–287.

Louchakova, O. (2007c). Spiritual heart and direct knowing in the Prayer of the Heart. *Existential Psychoanalysis, 18*(1), 81–102.

Louchakova, O. (2008). Reconstitution of the self during the study of Advaita Vedanta (Traditional Indian teachings of non-dual consciousness): Phenomenological investigations. In *Toward a science of consciousness,* April 8–12, 2008, Tucson AZ, Consciousness Research Abstracts (315), p. 194.

Louchakova-Schwartz, O. (2011). Self and world: Vedanta, Sufism and Presocratics in phenomenological view. In A.-T. Tymieniecka (Ed.), *Phenomenology/ontopoiesis retrieving cosmic horizons of antiquity* (pp. 423–438). Dordrecht, Germany: Springer.

Louchakova-Schwartz, O. (2012). Intuition of life in Tymieniecka's Phenomenology with a reference to intuition of Sat in Śakara's Advaita Vedanta. *Culture & Philosophy: A Journal for Phenomenological Inquiry, 2011*, 40–60.

Louchakova, O., & Warner, A. (2003). Via Kundalini: Psychosomatic excursions in transpersonal psychology. *The Humanistic Psychologist, 31*(2–3), 115–158.

Luck, S. J., & Vogel, E. K. (1997). The capacity of visual working memory for features and conjunctions. *Nature, 390,* 279–281.

Luft, Sebastian. 2011. *Subjectivity and lifeworld in transcendental phenomenology.* Evanston, Ill: Northwestern University Press.

Luhmann, N. (1990). The autopoiesis of social systems. In N. Luhmann (Ed.), *Essays on self reference*. New York, NY: Columbia University Press.

Luine, V. N., & McEwen, B. S. (1997). Effect of estradiol on turnover of the A monoamine oxidase in the brain. *Journal of Neurochemistry, 28,* 1221–1227.

Lundervold, A. (2010). On consciousness, resting state fMRI, and neurodynamics. *Nonlinear Biomedical Physics, 4* (Suppl. 1), S1–S9.

Luria, A. R. (1976). *Cognitive development: Its cultural and social foundations* (M. Cole, Ed., M. Lopez-Morillas, & L. Solotaroff, Trans.). Cambridge, MA: Harvard University Press.

Lutz, A., Brefczynski-Lewis, J., Johnstone, T., & Davidson, R. J. (2008). Regulation of the neural circuitry of emotion by compassion meditation: Effects of meditative expertise. *Public Library of Science One, 3* (3), e1897.

Lutz, A., Dunne, J. D., & Davidson, R. J. (2007). Meditation and the neuroscience of consciousness: An introduction. In P. Zelazo, M. Moscovitch, & E. Thompson (Eds.), *The Cambridge handbook of consciousness* (pp. 499–554). New York, NY: Cambridge University Press.

Lutz, A., Greischar, L. L., Rawlings, N. B., Richard, M., Davidson, R. J. (2004). Long-term meditators self-induce high-amplitude gamma synchrony during mental practice. *Proceedings of the National Academy of Sciences, 101*(46), 16369–16373.

Lutz, A., Lachaux, J.-P., Martinerie, J., & Varela, F. J. (2002). Guiding the study of brain dynamics using first-person data: Synchrony patterns correlate with on-going conscious states during a simple visual task. *Proceedings of the National Academy of Science USA, 99(3),* 1586–1591.

Lutz, A., Slagter, H. A., Dunne, J. D., & Davidson, R. J. (2008). Attention regulation and monitoring in meditation. *Trends in Cognitive Science, 12*(4), 163–169.

Lutz, A., & Thompson, E. (2003). Neurophenomenology: Integrating subjective experience and brain dynamics in the neuroscience of consciousness. *Journal of Consciousness Studies, 10*(9–10), 31–52.

MacKenzie, M. (2010). Enacting the self: Buddhist and enactive approaches to the emergence of the self. *Phenomenology and Cognitive Science, 9*(1), 75–99.

MacKenzie, M. (2012). Enacting the self: Buddhist and enactive approaches to the emergence of the self. In M. Sidertis, E. Thompson, and D. Zahavi (Eds.), *Self, no self? Perspectives from analytical phenomenological, & Indian traditions* (pp. 239–273). New York, NY: Oxford University Press. (Reprinted from 2010, *Phenomenology and the Cognitive Sciences, 9,* 75–99.)

MacKeracher, D. (2004). *Making sense of adult learning.* Toronto, ON: University of Toronto Press.

MacLean, P. D. (1949). Psychosomatic disease and the "visceral brain": Recent developments bearing on the Papez theory of emotion. *Psychosomatic Medicine, 11,* 338–353.

Majewska, M. D. (1987). Steroids and brain activity: Essential dialogue between body and mind. *Biochemical Pharmacology, 36,* 3781–3788.

Majewska, M. D., Harrison, N. L., Schwartz, R. D., Barker, J. L., & Paul, S. M. (1986). Steroid hormone metabolites are barbiturate-like modulators of the GABA receptor. *Science, 232,* 1004–1007.

Marbach, E. (2006). No heterophenomenology without autophenomenology: Variations on a theme of mine. *Phenomenology and the Cognitive Sciences, 6*(1–2), 75–87.

Marti S., Sackur, J., Sigman, M., & Dehaene, S. (2010). Mapping introspection's blind spot: reconstruction of dual-task phenomenology using quantified introspection. *Cognition, 115*(2), 303–313.

Martin, J., & Sugarman, J. (1997). Societal-psychological constructionism: Societies, selves, traditions and fusions. *Journal of Theoretical and Philosophical Psychology, 17*(2), 120–136.

Martinerie, J., Adams, C., Le Van Quyen, M., Baulac, M., Clémenceau, S., & Renault, B. (1998). Epileptic seizures can be anticipated by non-linear analysis, *Natural Medicine, 4,* 1173–1176.

Maslow, A. (1954). *Motivation and personality.* New York, NY: Harper and Row.

Maslow, A. (1971). A theory of metamotivation: The biological rooting of the value life. In A. Maslow (Ed.), *Farther reaches of human nature* (pp. 299–339). New York, NY: Viking. (Original work published 1967)

Maso, I. (2003). Necessary subjectivity: Exploiting researchers' motives, passions and prejudices in pursuit of answering 'true' questions. In L. Finlay & B. Gough (Eds.), *Reflexivity: A practical guide for researchers in health and social sciences.* Oxford, NY: Blackwell Publishing.

Maturana, H. R. (1975). The neuropsychology of cognition. In P. Garvin (Ed.), *Cognition: A multiple view.* New York, NY: Spartan.

Maturana, H. R. (1988a). Reality: The search for objectivity or the quest for a compelling argument. *Irish Journal of Psychology, 9,* 25–82

Maturana, H. R. (1988b). Ontology of observing: The biological foundations of self, consciousness and the physical domain of existence. In R. Donaldson (Ed.), *Texts in cybernetic theory: An in-depth exploration of the thought of Humberto Maturana, William T. Powers, and Ernst von Glaserfeld.* Felton, CA: American Society for Cybernetics. Retrieved 9/4/12 from http://www.inteco.sl/biology/ontology/index.htm.

Maturana, H. R., & Varela, F. J. (1973), De máquinas y seres vivos: Una teoría de la organizacíon biológica (Santiago, Chile: Editorial Univeristaria).

Maturana, H. R., & Varela, F. J. (1980). *Autopoiesis and cognition: The realization of the Living. Boston Studies in the Philosophy of Science, 42.* Boston, MA; Dordrecht, Holland: D. Reidel.

Maturana, H., & Varela, F. (1987). *The tree of knowledge: The biological roots of human understanding.* Boston, MA: New Science Library, Shambhala Publications.

Maturana, H. R., Varela, F. J., & Uribe, R. (1974). Autopoiesis: The organization of living systems, its characterization and a model. *Biosystems, 5,* 187–196.

Mauss, I. B., Levenson, R.W., McCarter, L., Wilhelm, F. H., & Gross, J. J. (2005). The tie that binds? Coherence among emotion experience, behavior, and physiology. *Emotion, 5,* 175–190.

May, R. (Ed.). (1961). *Existential psychology.* New York, NY: Random House.

May, R. (1975). *The courage to create.* New York, NY: Norton.

May, R., Angel, E., & Ellenberger, H. F. (Eds.). (1958). *Existence: A new dimension in psychiatry and psychology.* New York, NY: Basic Books.

Mayo, K. R. (2009). Support from neurobiology for spiritual techniques for anxiety: A brief review. *Journal of Health Care Chaplaincy, 16*(1–2), 53–57.

McDermott J. J. (Ed.). (1977). *Writings of William James: A comprehensive edition.* Chicago, IL: University of Chicago Press.

McGaugh, J. L. (2000). Memory – a century of consolidation. *Science, 287,* 248–251.

McGaugh, J. L. (2004). The amygdala modulates the consolidation of memories and emotionally arousing experiences. *Annual Review of Neuroscience, 27,* 1–28.

McInerney, R. G. (2010). Oz never did give nothing to the… Scarecrow: Neurophenomenology and critical pedagogy. *Phenomenology and Practice, 4*(1), 68–87.

Megill, A. (1994). Four senses of objectivity. In A. Megill (Ed.), *Rethinking objectivity* (pp. 37–64). Durham, NC; London, UK: Duke University Press.

Melzoff, A., & Moore, M. K. (1977). Imitation of facial and manual gestures by human neonates. *Science, 198,* 75–78.

Melzoff, A., & Moore, M. K. (1983). Newborn infants imitate adult facial gestures. *Child Development, 54,* 702–709.

Melzoff, A., & Moore, M. K. (1989). Imitation in newborn infants: Exploring the range of gestures imitated and the underlying mechanisms. *Developmental Psychology, 25,* 954–962.

Mennes, M., Zuo, X. N., Kelly, C., Di Martino, A., Zang, Y. F., Biswal, B.,…Milham, M. P. (2011). Linking inter-individual differences in neural activation and behavior to intrinsic brain dynamics. *Neuroimage, 54*(4), 2950–2959.

Merleau-Ponty, M. (1962). *Phenomenology of perception* (C. Smith, Trans.). London, UK; New York, NY: Routledge and Kegan Paul. (Original work published 1945)

Merleau-Ponty, M. (1964a). *Le Visible et l'Invisible.* Paris, France: Galland.

Merleau-Ponty, M. (1964b). *The primacy of perception.* (J. M. Edie, Ed.). Evanston, IL: Northwestern University Press.

Merleau-Ponty, M. (1968). *The visible and the invisible.* (A. Lingis, Trans.). Evanston, IL: Northwestern University Press. (Original work published 1964)

Merleau-Ponty, M. (1983a). *The Structure of behavior.* Pittsburgh, PA: Duquesne University Press. (Original work published 1947)

Merleau-Ponty, M. (1983b). *The structure of behavior* (A. L. Fisher, Trans.). Pittsburgh, PA: Duquesne University Press. (Original work published 1942)

Merleau-Ponty, M. (1989). *Phenomenology of perception* (C. Smith, Trans.). London, UK: Routledge. (Original work published 1962)

Merleau-Ponty, M. (2002). *Phenomenology of perception* (2nd ed.). New York, NY: Routledge.

Metzler, J., & Shepard, R. N. (1974). Transformational studies of the internal representation of three-dimensional objects. In R. Solso (Ed.), *Theories in cognitive psychology: The Loyola Symposium* (pp. 147–201). Hillsdale, NJ: Erlbaum.

Mezirow, J. (1991). *Transformative dimensions of adult learning.* San Francisco, CA: Jossey-Bass.

Mitchel, D. J., & Cusack, R. (2011). The temporal evolution of electromagnetic markers sensitive to the capacity limits of visual short-term memory. *Frontiers in Human Neuroscience, 5,* 18.

Morey, C. C., Cowan, N., Morey, R. D., & Rouder, J. N. (2011). Flexible attention allocation to visual and auditory working memory tasks: Manipulating reward induces a trade-off. *Attention, Perception, & Psychophysics, 3*(2), 458–472.

Moriguchi, Y., Ohnishi, T., Mori, T., Matsuda, H., & Komaki, G. (2007). Changes of brain activity in the neural substrates for theory of mind during childhood and adolescence. *Psychiatry in Clinical Neuroscience, 61* (4), 355–363.

Moustakas, C. E. (1961). *Loneliness.* Englewood Cliffs, NJ: Prentice-Hall.

Moustakas, C. E. (1994). *Phenomenological research methods.* New York, NY: Sage Publications.

Mruk, C. J. (2008). The psychology of self-esteem: A potential common ground for humanistic positive psychology and positivistic positive psychology. *The Humanistic Psychologist, 36*(2), 143–158.

Murphy, N. (1999). Introduction. In R. Russel, N. Murphy, T. Meyering, & M. Arbib (Eds.), *Neuroscience and the person. Scientific perspectives on divine action* (pp. i-xxxv). Berkeley, CA: Center for Theology and Natural Science.

Myers, F. W. H. (1886). Multiplex personality. *Proceedings of the English Society for Psychical Research, 4,* 496–514.

Myers, F. W. H. (1892).The subliminal consciousness: Chapter 1. General characteristics of subliminal messages. *Proceedings of the Society for Psychical Research, 7,* 298–327.

Myers, F. W. H. (2001). *Human personality and its survival of bodily death.* Charlottesville, VA: Hampton Roads. (Original work published 1903)

Nagel, T. (1971). Brain bisection and the unity of consciousness. *Synthese, 22,* 396–413.

Neisser, U. (1976). *Cognition and reality: Principles and implications of cognitive psychology.* New York, NY: W. H. Freeman & Co.

Neumann, N.-U., & Karel, F. (2006). The neurobiological dimension of meditation - results from neuroimaging studies. *Psychotherapie Psychosomatik Medizinische Psychologie, 5* (12), 488–492.

Newberg, A. B. (2006). Religious and spiritual practices: A neurochemical perspective. In P. NcNamara (Ed.), *Where God and science meet: How brain and evolutionary studies alter our understanding of religion* (pp. 15–32). Westport, CT: Praeger.

Newberg, A. B. (2010). Neurobiology of meditation. In D. Monty & B. Beitman (Eds.), *Integrative psychiatry* (pp. 339–358). Oxford, NY: Oxford University Press.

Newberg, A., Alavi, A., Baime, M., Pourdehnad, M., Santanna, J., & d'Aquili, E. (2001). The measurement of regional cerebral blood flow during the complex cognitive task of meditation: A preliminary SPECT study. *Psychiatry Research, 106,* 113–122.

Newberg, A., Pourdehnad, M., Alavi, A., & d'Aquili, E. G. (2003). Cerebral blood flow during meditative prayer: Preliminary findings and methodological issues. *Perceptual and Motor Skills, 97,* 625–630.

Noë, A. (2004). *Action in perception.* Cambridge, MA: The MIT Press.

Noe, A. (2006). *Action in perception.* Cambridge, MA: The MIT Press.

Noë A. (2009). *Out of our heads: Why you are not your brain, and other lessons from the biology of consciousness.* New York, NY: Hill & Wang.

Noe, A. (2010). *Out of our heads: Why you are not your brain, and other lessons from the biology of* consciousness. New York, NY: Hill & Wang.

Notterman, J. M. (2000). Note on reductionism in cognitive psychology: Reification of cognitive processes in to the mind, mind-brain equivalence and brain-computer analogy. *Journal of Theoretical and Philosophical Psychology, 20,* 116–121.

Ochsner, K. N., Bunge, S. A., Gross, J. J., & Gabrieli, J. D. (2002). Rethinking feelings: An fMRI study of the cognitive regulation of emotion. *Journal of Cognitive Neuroscience 14*(8), 1215–1229.

Ochsner, K. N., Ray, R. D., Cooper, J. C., Robertson, E. R., Chopra, S., & Gabrieli, J. D. E,… Gross, J. J. (2004). For better or worse: Neural system supporting the cognitive down-and upregulation of negative emotion. *NeuroImage, 23*(2), 483–499.

Ochsner, K. N., Ray. R. R., Hughes, B., McCrae, K., Cooper, J. C., Weber, J.,…Gross, J. J. (2009). Bottom up and top down processes in emotion generation: Common and distinct central mechanism. *Psychological Science, 20*(11), 1322–13331.

Osbeck, L. M. (2009). Transformations in cognitive science: Implications and issues posed. *Journal of Theoretical and Philosophical Psychology, 29*(1), 16–33.

Ostrosky, M. M., & Kaiser, A. P. (1991). Preschool classroom environments that promote communication. *Teaching Exceptional Children, 2*(34), 6–10.

Overgaard, M. (2004). On the naturalising of phenomenology. *Phenomenology and the Cognitive Sciences, 3,* 365–379.

Pagnoni, G., Cekic, M., & Guo, Y. (2008). Thinking about not-thinking: Neural correlates of conceptual processing during Zen meditation. *Public Library of Science One, 3*(9), e3083.

Papez, J. W. (1937). A proposal mechanism of emotion. *Journal of Neuropsychiatry and Clinical Neuroscience, 7*(1), 103–112; *Archives of Neurological Psychiatry, 38, 725*–743.

Paré, D., Collins, D. R., & Pelletier, J. G. (2002), Amygdala oscillations and the consolidation of emotional memories. *Trends in Cognitive Sciences, 6,* 306–14.

Paulson, L., & Paulson, P. (1994). Assessing portfolios using the constructivist paradigm. In R. Fogarty (Ed.), *Student portfolios.* Palatine, IL: IRI Skylight Training & Publishing.

Pavlov, I. (1960). *Conditioned reflexes: An introduction of the physiological activity of the cerebral cortex* (G.V. Anrep, Ed., and Trans.). New York, NY: Dover. (Original work published 1927)

Pea, R. (1993). Practices of distributed intelligence and designs for education. In G. Salomon (Ed.), *Distributed cognitions: Psychological and educational considerations.* Cambridge, MA: Cambridge University Press.

Pekala, R., Wenger, C., & Levine, R. (1985). Individual differences in phenomenological experience: States of consciousness as a function of absorption. *Journal of Personality and Social Psychology, 48*(1), 125–32.

Pert, C. B. (1999). *Molecules of emotion.* New York, NY: Touchstone.

Petitmengin, C. (Ed.). (2009). *Ten years of viewing from within: The legacy of Francisco Varela.* Thorverton, UK: Imprint Academic.

Petitmengin, C., & Bitbol, M. (2009). The validity of first-person descriptions as authenticity and coherence. In C. Petitmengin (Ed.), *Ten years of viewing from within: The legacy of Francisco Varela* (pp. 363–404). London, UK: Imprint Academic

Petitot, J. (1992). *Physique du sens.* Paris, France: Editions du CNRS.

Petitot, J., Varela, F. J., Pachoud, B., & Roy, J.-M. (Eds.). (1999). *Naturalizing phenomenology: Issues in contemporary phenomenology and cognitive science.* Stanford, CA: Stanford University Press.

Pfeiffer, W. M. (1966). Concentrative self-relaxation by exercises derived from Buddhistic respiration- meditation and from respiration therapy. Zeitschrift für Psychotherapie und *Medizinische Psychologie, 16*(5), 172–81.

Phelps, E. A. (2006). Emotion and cognition: Insights from studies of the human amygdala. *Annual review of Psychology, 57,* 27–53.

Phelps, E. A., & LeDoux, J. E. (2005). Contributions of the amygdala to emotion processing: From animal models to human behavior. *Neuron, 48,* 175–187.

Phelps, E. A., Ling, S., & Carrasco, M. (2006). Emotion facilitates perception and potentiate the perceptual benefits of attention. *Psychological Science, 17*(4), 292–299.

Piaget, J. (1975). *The development of thought: Equilibrium of cognitive structures.* New York, NY: Viking Press.

Pinar, W. F. (2004). *What is curriculum theory?* Mahwah, NJ: Lawrence, Erlbaum.

Poerksen, B. (2004). *The certainty of uncertainty: Dialogues introducing constructivism* (A. R. Koeck, & W. F. Koeck, Trans.). London, UK: Imprint Academic.

Polanyi, M. (1966). *The tacit dimension.* New York, NY: Doubleday.

Poremba, A., & Gabriel, M. (2001). Amygdalar efferents initiate auditory thalamic discriminative training-induced neuronal activity. *Journal of Neuroscience, 21,* 260–268.

Posner, M. I., & Dehaene, S. (1994). Attentional networks. *Trends in Neuroscience, 17,* 75–79.

Posner, M. I., & Petersen, S. E. (1990). The attentional system of the human brain. *Annual Review of Neuroscience, 13*, 25–42.

Quirk, G. J., Armony, J. L., & LeDoux, J. E. (1997). Fear conditioning enhances difference temporal components of tone-evoked spike trains in auditory cortex and lateral amygdala. *Neuron, 19,* 613–624.

Radley, J. J., Arias, C. M., & Sawchenko, P. E. (2006). Regional differentiation of the medial prefrontal cortex in regulating adaptive responses to acute emotional stress. *The Journal of Neuroscience, 26*(50), 12967–12976.

Raffone, A., & Pantani, M. (2010). A global workspace model for phenomenal and access consciousness. *Consciousness and Cognition, 19*(2), 580–596.

Raffone, A., & Srinivasan, N. (2009). An adaptive workspace hypothesis about the neural correlates of consciousness: Insights from neuroscience and meditation studies. *Progressive Brain Research, 176,* 161–180.

Raffone, A., & Srinivasan, N. (2010). The exploration of meditation in the neuroscience of attention and consciousness. *Cognitive Processing, 11*(1), 1–7.

Rapkin, A. J. (1999). Progesterone, GABA and mood disorders in women. *Archives of Women's Health, 2,* 97–105.

Re, R. N. (2002). The origins of intracrine hormone action. *American Journal of Medical Science, 323*(1), 43–48.

Re, R. N. (2003). The intracrine hypothesis and intracellular peptide hormone action. *BioEssays, 25,* 401–409.

Re, R. N., & Bryan, S. E. (1984). Functional intracellular rennin-angiotensin systems may exist in multiple tissues. *Hypertension Clinical Experimental Theory Tract, A6* (Suppl. 10 & 11), 1739–1742.

Re, R. N., & Cook, J. L. (2006). The intracrine hypothesis: An update. *Regulatory Peptide, 133*(1–3), 1–9.

Re, R. N., & Cook, J. L. (2008). The basis of an intracrine pharmacology. *Journal of Clinical Pharmacology, 48*(3), 344–350.

Rietveld, E. (2008). The skillful body as concernful system of possible actions: Phenomena and neurodynamics. *Theory & Psychology, 18*(3), 341–364.

Risold, P. Y. & Swanson, L. W. (1996). Structural evidence for functional domains in the rat hippocampus. *Science, 272,* 1484–1486.

Rizzolatti, G., Fadiga, L., Gallese, V., & Fogassi, L. (1996). Premotor cortex and the recognition of motor actions. *Cognitive Brain Research, 3,* 131–141.

Robbins, B. D. (2003). *Joy and the politics of emotion: Towards a cultural therapeutics via phenomenology and critical theory*. Doctoral dissertation, Duquesne University, Pittsburgh, PA.

Robbins, B. D. (2006). An empirical, phenomenological study: Being joyful. In C. T. Fischer (Ed.), *Qualitative research methods for psychologists: Introduction through empirical studies* (pp. 173–212). Burlington, MA: Elsevier, Academic Press.

Robbins, B. D., & Fleming, R. (2008, August). *The unwanted exposure model of embarrassment: Experimental validation of an empirical, phenomenological study.* Poster presented at the 2nd Annual Conference of the Society for Humanistic Psychology, Division 32 of APA, Norwood, MA.

Robbins, B. D., & Parlavecchio, H. (2006). The unwanted exposure of the self: A phenomenological study of embarrassment. *The Humanistic Psychologist, 34*(4), 321–345.

Robbins, B. D., & Vandree, K. (2009). The self-regulation of humor expression: A mixed method, phenomenological investigation of suppressed laughter. *The Humanistic Psychologist, 3*(1), 49–78.

Robbins, B. D., & Gordon, S. (in press). Humanistic neuropsychology: The implications of neurophenomenology for psychology. In K. J. Schneider & J. F. Pierson (Eds.), *The handbook of humanistic psychology: Leading edges in theory, research, and practice* (2nd ed.). Thousand Oaks, CA: Sage Publications.

Robel, P., & Baulieu, E. E. (1995). Dehydroepiandrosterone (DHEA) is a neuroactive steroid. *Annals of the New York Academy of Science, 774,* 82–110.

Rodriguez, E., Lachaux, G. N., Martinerie, J., Renault, B., & Varela, F. J. (1999). Perception's shadow: Long-distance synchronization in the human brain. *Nature 397,* 340–343.

Rogers, C. (1964). Toward a science of the person. In T. W. Wann (Ed.), *Behaviorism and phenomenology: Contrasting bases for modern psychology* (pp. 109–140). Chicago, IL: University of Chicago Press.

Rose, S. (2005). *The future of the brain: The promise and perils of tomorrow's neuroscience.* New York, NY: Oxford University Press.

Rosenzweig, M. R. (2003). Effects of differential experience on the brain and behavior. *Developmental Neuropsychology, 24*(2&3), 523–540.

Roy, M-J, Petitot, J., Pachoud, B., & Varela, F. J. (1999). Beyond the gap: An introduction to naturalizing phenomenology. In J. Petitot, F. J. Varela, B. Pachoud & J-M, Roy (Eds.), *Naturalizing phenomenology: Issues in contemporary phenomenology and cognitive science* (pp. 1–82). Stanford, CA: Stanford University Press.

Rubik, B. (2011). Neurofeedback-enhanced gamma brainwaves from the prefrontal cortical region of meditators and non-meditators and associated subjective experiences. *Journal of Alternative and Complementary Medicine, 17*(2), 109–115.

Rudrauf, D., Lutz, A., Cosmelli, D., Lachaux, J.-P., & Le Van Quyen, M. (2003). From autopoiesis to neurophenomenology: Francisco Varela's exploration of the biophysics of being. *Biological Research, 36,* 21–59; 27–65.

Rupprecht, R., Koch, M., Montkowski, A., Lancel, M., Faulhaber, J., & Harting, J.,…Spanagel, R. (1999). Assessment of neuroleptic-like properties of progesterone *Psychopharmacology, 143*(1), 29–38.

Ryan, J., & Herman, H. P. (2008). Limbic regulation of hypothalamo-pituitary-adrenocortical function during acute and chronic stress. *Annals of the New York Academy of Science, 1148,* 64–73.

Sacks, P. (1999). *Standardized minds: The high price of America's testing culture and what we can do to change it.* Cambridge, MA: Perseus Books.

Sanders, D., Warner, P., Backstrom, T., & Bancroft, J. (1983). Mood, sexuality, hormones and the menstrual cycle: Change in mood and physical state: Description of subjects and method *Psychosomatic Medicine, 45*(6), 487–501.

Sansonese, J. N. (1994). *The body of myth: Mythology, shamanic trace, and the sacred geography of the body.* Rochester, VT: Inner Traditions International.

Sartre, J. P. (2003). *Sketch for a theory of the emotions* (2nd ed.). New York, NY: Routledge. (Original work published 1939)

Sawyer, R. K., & Greeno, J. G. (2009). Situativity and learning. In P. Robbins & M. Aydede (Eds.), *The Cambridge handbook of situated cognition.* Cambridge, MA: Cambridge University Press.

Schiff, S. J. (1998). Forecasting brain storms. *Nature Medicine, 4*(10), 1117–1118.

Schore, A. (2010). The right brain implicit self: A central mechanism of the psychotherapy change process. In J. Petrucelli (Ed.), *Knowing, not-knowing and sort of knowing: Psychoanalysis and the experience of uncertainty* (pp. 177–202). London, UK: Karnac.

Schwartz, G. E., Davidson, R. J., & Goleman, D. J. (1978). Patterning of cognitive and somatic processes in the self-regulation of anxiety: Effects of meditation versus exercise. *Psychosomatic Medicine, 40*(4), 321–8.

Searle, J. (2005). The self as a problem in philosophy and neurobiology. In T. E. Feinberg & J. P. Keenan (Eds.), *The lost self: Pathologies of the brain and identity.* Oxford, NY: Oxford University Press.

Sells, M. A. (1994). *Mystical languages of the unsaying.* Chicago, IL: University of Chicago Press.

Selye, H. (1946). Correlations between the chemical structure and the pharmacologic actions of steroids. *Endocrinology, 30,* 437–453.

Shaoyu, G., Chih-hao, Y., Kuei-sen, H., Guo-li, M., & Hongjun, S. (2007). A critical period for enhanced synaptic plasticity in newly generated neurons of the adult brain. *Neuron, 54*(4), 559–56.

Sherwin, B. B., & Surayani-Cadotte, B. E. (1990). Up-regulatory effect of estrogen on platelet 3H-impramine binding sites in surgically menopausal women. *Biological Psychiatry, 28,* 339–348.

Sheets-Johnstone, M. (1990). *The roots of thinking.* Philadelphia, PA: Temple University Press.
Siderits, M., Thompson, E., & Zahavi, D. (Eds.). (2011). *Self, no self? Perspectives from analytical, phenomenological, and Indian traditions.* New York, NY: Oxford University Press.
Sidis, B. (1898). *The psychology of suggestion.* New York, NY: Appleton and Co.
Siegel, D. (1999). *The developing mind: How relationships and the brain interact to shape who we are.* New York, NY: Guilford Press.
Siegel, D. J. (2010). *The mindful therapist: A clinician's guide to mindsight and neural integration*. New York, NY: W.W. Norton and Co.
Simms, E. M. (1999). The countryside of childhood: A hermeneutic phenomenological approach to developmental psychology. *The Humanistic Psychologist, 27*(3), 301–328.
Simpson, E. R. (2002). Aromatization of androgens in women: Current concept and findings. *Fertility and Sterility, 77*(4), S6-S10.
Skene, J. H. (1989). Axonal growth-associated proteins. *Annual Review of Neuroscience, 12,* 127–156.
Skinner, B. F. (1953). *Science and human behavior.* New York, NY: Macmillan Press.
Skinner, B. F. (1974). *About behaviorism.* New York, NY: Alfred A. Knopf.
Slagter, H. A., Davidson, R. J., & Lutz, A. (2011). Mental training as a tool in the neuroscientific study of brain and cognitive plasticity. *Frontiers in Human Neuroscience, 10,* 5–17.
Sousa, D. A. (2010). *Mind, brain and education. Neuroscience implications for the classroom.* Bloomington, IN: Solution Tree Press.
Spinelli, E. (1989). *The interpreted world: An introduction to phenomenological psychology.* London, UK: Sage.
Squire, L. R., Berg, D., Bloom, F. E., Du lac, S., Ghosh, A., & Spitzer N. C. (2008). *Fundamental neuroscience* (3rd ed.). Burlington, MA: Elsevier Inc.
Stcherbatskoi (Stcherbatsky, Sherbastky), T. (1903). *Teoriya poznaniya i logika po ucheniyu pozdneishikh buddistov* [*Theory of knowledge and Logic According to Later Buddhists*]. St. Petersburg, Russia: Izd. Fak. Vost Yazykov Imp. Sankt-Peterburgskogo Universiteta. No. 14.
Strasser, S. (1967). Phenomenologies and psychologies. In N. Lawrence & D. O'Connor (Eds.), *Reading in existential-phenomenology.* Englewood Cliffs, NJ: Prentice-Hall.
Strauss, J. F., & Barbieri, R. L. (Eds.). (2009). *Yen and Jaffe's reproductive endocrinology: Physiology, pathophysiology, and clinical management* (6th ed.). Philadelphia, PA: W. B. Saunders.
Szasz, T. (1996). *The meaning of mind: Language, morality, and neuroscience.* Westport, CT: Praeger Publishing.
Taves, A. (1999). *Fits, trances, and visions: Experiencing religion and explaining experience from Wesley to James.* Princeton, NJ: Princeton University Press.
Taylor, E. I. (1973). *Psychological suspended animation: Heart rate, blood pressure, time estimation, and introspective reports from an anechoic environment.* Master of Arts Thesis, Department of Psychology, Southern Methodist University, Dallas, TX. Privately printed by The Essene Press.
Taylor, E. I. (1982). *William James on exceptional mental states: Reconstruction of the 1896 Lowell lectures.* New York, NY: Charles Scribner's Sons; reproduced in paperback by the University of Massachusetts Press, Amherst, 1984.
Taylor, E. I. (1984). *William James on exceptional mental states: The 1896 Lowell Lectures.* Amherst, MA: The University of Massachusetts Press.
Taylor, E. I. (1990). William James on Darwin: An evolutionary theory of consciousness. *Annals of the New York Academy of Sciences, 602,* 7–33. Originally presented as part of the Darwin Centennial Lectures.
Taylor, E. I. (1992a). *Psychology as a person-centered science: William James after 1890.* (Doctoral dissertation, Boston University). *Dissertation Abstracts International, B 53/03.* (UMI No. 9222837).
Taylor, E. I. (1992b). The case for a uniquely American Jamesian tradition in psychology. In M. Donnelly (Ed.), *Reinterpreting the legacy of William James.* (APA Centennial William James Lectures). (pp. 3–28).Washington, D.C.: American Psychological Association.

Taylor, E. I. (1996). *William James on consciousness beyond the margin*. Princeton, NJ: Princeton University Press.

Taylor, E. I. (1997). *A psychology of spiritual healing*. West Chester, PA: Chrysalis Books.

Taylor, E. I. (1999). *Shadow culture: Psychology and spirituality in America.* Washington, D.C.: Counterpoint.

Taylor, E. I. (2000). "What is man, Psychologist, that thou art so unmindful of him?": Henry A, Murray on the historical relation between classical personality theory and humanistic psychology. *Journal of Humanistic Psychology, 40*(3), 29–42.

Taylor, E. I. (2001–2002). *William James and the spiritual roots of American pragmatism.* Lectures commemorating the Centenary of William James's *Varieties of Religious Experience,* delivered at Harvard University. Manuscript completed. Unpublished.

Taylor, E. I. (2003a). *Why Wundt's laboratory was not founded until 1950, Or the hypothesis of the three streams.* The Wallace Russell Memorial Lecture presented to (Division 26) The History of Psychology at the 109th annual meeting of the American Psychological Association, Toronto, Canada, August 9, 2003.

Taylor, E. I. (2003b). Have we engaged in a colossal misreading of James's *Varieties? Streams of William James*. Newsletter/journal of the William James Society.

Taylor, E. I. (2009). *The mystery of personality: A history of psychodynamic theories.* New York, NY: Springer.

Taylor, E. I. (2010a). *William James on exceptional mental states: Reconstruction of the 1896 unpublished Lowell lectures* (2nd printing). Portsmouth, NH: Jetty Press. (Original work published 1982)

Taylor, E. I. (2010b). William James on an intuitive, phenomenologically oriented psychology in the immediate moment: The true foundation for a science of consciousness? Special issue on Consciousness in the Neurosciences. *History of the Human Sciences, 23*(3), 119–130.

Taylor, E. I. (2010c). William James and the humanistic implications of the neuroscience revolution: An outrageous hypothesis. *Journal of Humanistic Psychology, 50*(4), 410–429.

Taylor, E. I. (2012a). The ontological triad in James and Peirce: A moment in the history of consciousness, *Revue Internationale de Philosophie, 260,* 177–186.

Taylor, E. I. (2012b). The stream of consciousness: Literary psychology as the first uniquely American phenomenology in the works of William James and his Swedenborgian and Transcendentalist milieu. In A.T. Tymieniecka (Ed.), *Art, Literature and Passions of the Skies, Analecta Husserliana, 108,* 112.

Taylor, E. I., Martinez, S., & Martin, F. (Eds.). (2000). *Old Saybrook One: Benchmark in the history of humanistic psychology. A commemorative edition.* Cambridge, MA: The Essene Press.

Taylor, J. G., & Fragopanagos, N. (2007). Resolving some confusions over attention and consciousness. *Neural Networks, 20*(9), 993–1003.

Tei, S., Faber, T. L., Lehmann, D., Tsujiuchi, T., Kumano, H., Pascual-Marqui, R. D.,…Kochi, K. (2009). Meditators and non-meditators: EEG source imaging during resting. *Brain Topography, 22*(3), 158–165.

Thelen, E., & Smith, L. (1994). *A dynamic systems approach to the development of cognition and action*. Cambridge, MA: The MIT Press.

Thibault, P. J. (2004). *Brain, mind, and the signifying body: An ecosocial semiotic theory.* New York, NY: Continuum Publishing.

Thom, R. (1975). *Structural stability and morphogenesis.* Reading, MA: W. A. Benjamin.

Thomas, N. J. T. (2010). Mental imagery. In E. N. Zalta (Ed.), *The Stanford encyclopedia of philosophy.* Retrieved from http://plato.stanford.edu/archives/fall2010/entries/mental-imagery/

Thompson, E. (Ed.). (2001a). *Between ourselves: Second-person issues in the study of consciousness*. Charlottesville, VA: Imprint Academic.

Thompson, E. (2001b). Obituary of Francisco J. Varela (1946–2001). *Journal of Consciousness Studies, 8,* 66–69). See also (http://psyche.cs.monash.edu.au/v7/psyche-7-12-thompson.html

Thompson, E. (2005). Sensorimotor subjectivity and the enactive approach to experience. *Phenomenology and the Cognitive Sciences, 4*(4), 407–427.

Thompson, E. (2007). *Mind in life: Biology, phenomenology, and the sciences of mind.* Cambridge, MA: Harvard University Press.

Thompson, E., Lutz, A., & Cosmelli, D. (2005). Neurophenomenology: An introduction for neurophilosophers. In A. Brook & K. Akins (Eds.), *Cognition and the brain: The philosophy and neuroscience movement* (pp. 40–97). New York, NY: Cambridge University Press.

Thompson, E., & Varela, F. J. (2001). Radical embodiment. *Trends in Cognitive Science 5*(10), 418–425.

Thompson, J., Maddock, M., Ayaland, M., & Oswald, I. (1977). Relationship between nocturnal plasma oestrogen concentration and free plasma tryptophan in perimenopausal women. *Journal of Endocrinology, 72* , 395–396.

Thorndike, E. L. (1931). *Human learning.* New York, NY: Appleton.

Toadvine, T. (2002). Leaving Husserl's cave? The philosopher's shadow revisited. In T. Toadvine, & L. Embree (Eds.), *Merleau-Ponty's readings of Husserl* (pp. 71–94). Netherlands: Kluwer.

Tolman, E. C. (1922). A new formula for behaviorism. *Psychological Review, 29,* 44–53.

Tolman, E. C. (1948). Cognitive maps in rats and men. *Psychological Review, 55,* 189–208.

Toren, C. (2001). The child in mind. In H. Whitehouse (Ed.), *The debated mind: Evolutionary psychology versus ethnography* (pp. 155–179). New York, NY: Berg, Oxford International Press.

Travis, F., Arenander, A., & DuBois, D. (2004). Psychological and physiological characteristics of a proposed object-referral/self-referral continuum of self-awareness. *Consciousness and Cognition, 13,* 401–420.

Travis, T., & Shear, J. (2010). Focused attention, open monitoring and automatic self-transcending: Categories to organize meditations from Vedic, Buddhist and Chinese traditions. *Consciousness and Cognition, 19*(4), 1110–1118.

Travis, F., & Wallace, R. K. (1999). Autonomic and EEG patterns during eyes-closed rest and transcendental meditation (TM) practice: The basis for a neural model of TM practice. *Conscious and Cognition, 8*(3), 302–318.

Tsigos, C., & Chrousos, G. P. (2002). Hypothalamic-pituitary-adrenal axis, neuroendocrine factors and stress. *Journal of Psychosomatic Research, 53*(4), 865–871.

Tymienieka, A.T. (1989). The theme: The history of American phenomenology-in-process. In E. F. Kaelin & C. O. Schrag (Eds.), *American phenomenology: Origins and development, Analecta Hussereliana, 26,* (ix-xxvii). Brattleboro, VT: World Institute for Advanced Phenomenological Research and Learning.

Uhlhaas, P. J., & Mishara, A. L. (2007). Perceptual anomalies in schizophrenia: Integrating phenomenology and cognitive neuroscience. *Schizophrenia Bulletin, 33*(1), 142–156.

Uttal, W. R. (2011). *Mind and brain: A critical appraisal of cognitive neuroscience.* Cambridge, MA: The MIT Press.

Van den Heuvel, M. P., & Hulshoff Pol, H. E. (2010). Exploring the brain network: A review on resting-state fMRI functional connectivity. *European Neuropsychopharmacology, 20*(8), 519–534.

Vanderstraeten, R. (2002). Dewey's transactional constructivism. *Journal of Philosophy of Education, 36*(2), 233–246.

van Gelder, T. J. (1998). The dynamical hypothesis in cognitive science. *Behavioral and Brain Sciences, 2,* 1–14.

van Manen, M. (1990). *Researching lived experience: Human science for an action sensitive pedagogy.* New York, NY: State University of New York Press.

Varela, F. J. (1979a). *The principles of biological autonomy.* New York, NY: Oxford and Elsevier, North Holland Inc.

Varela, F. J. (1979b). Reflections on the Chilean civil war. *Lindisfarne Letter, 8*(winter), 13–19.

Varela, F. J. (1981). Describing the logic of the living. In M. Zeleny (Eds.), *Autopoiesis: A theory of living organization* (pp. 36–48). New York, NY: Elsevier.

Varela, F. J. (1991). Organism: A meshwork of selfless selves. In A. I. Tauber (Ed.), *Organism and the origins of self* (pp. 79–107). Netherlands: Kluwer Academic Publishers.

Varela, F. J. (1992/1999). *Ethnical know-how: Action, wisdom and cognition.* Stanford, CA: Stanford University Press.
Varela, F. J. (1996). Neurophenomenology: A methodological remedy for the hard problem. *Journal of Consciousness Studies, 3*(4), 330–349.
Varela, F. J. (Ed.). (1997). *Sleeping, dreaming, and dying: An exploration of consciousness with the Dalai Lama.* Boston, MA: Wisdom Publications.
Varela, F. J. (1999). The specious present: A neurophenomenology of time consciousness. In J. Petitot, F. J. Varela., B. Pachoud, & J-M. Roy (Eds.), *Naturalizing phenomenology: Issues in contemporary phenomenology and cognitive science* (pp. 266–314). Stanford, CA: Stanford University Press.
Varela, F. J. (2001). Intimate distances: Fragments for a phenomenology of organ Transplantation. *Journal of Consciousness Studies, 8* (5–7), 259–71.
Varela, F. J., & Cohen, A. (1989). Le corps evocateur: Une relectre de L'immunité. *Novelle Revue de Psychanalyse, 40,* 193–213.
Varela, F. J., & Depraz, N. (2000). At the source of time: Valence and the constitutional dynamics of affect. *Journal de lettre et de sciences humain 4*(1–2).
Varela, F. J., & Goguen, J. (1977). The arithmetic of closure. In R. Trappel (Ed.), *Progress in cybernetics and systems research* (Vol. 3, pp. 48–63). New York, NY: Wiley Hemisphere.
Varela, F. J., Lachaux, J-P., Rodríguez, E., & Martinerie, J. (2001). The brainweb: Phase synchronization and large-scale integration. *Nature Reviews Neuroscience, 2,* 229–239.
Varela, F. J., & Maturana, H. (1973). Mechanism and biological explanation. *Philosophy of Science, 39,* 378–382.
Varela, F. J., & Shear, J. (Eds.). (1999a). *The view from within: First-person approaches to the study of consciousness.* Thorverton, UK: Imprint Academic.
Varela, F. J., & Shear, J. (1999b). First-person methodologies: What, why, how? *Journal of Consciousness Studies, 6*(2–3), 1–14.
Varela, F. J., & Thompson, E. (2003). Neural synchrony and the unity of mind: A neurophenomenological perspective. In A. Cleeremans (Ed.), *The unity of consciousness: Binding, integration and dissociation* (pp. 266–287). New York NY: Oxford University Press.
Varela, F. J., Thompson, E., & Rosch, E. (1991). *The embodied mind: Cognitive science and human experience* . Cambridge, MA: The MIT Press.
Vaz, N. M., & Varela, F. J. (1978). Self and non-sense: An organism-centered approach to immunology. *Medical Hypothesis, 4*(3), 231–261; 263–267.
Vogel, A. C., Power, J. D., Petersen, S. E., & Schlaggar, B. L. (2010). Development of the brain's functional network architecture. *Neuropsychology Review, 20*(4), 362–375.
Vogeley, K., May, M., Ritzl, A., Falkai, P., Zilles, K., & Fink, G. R. (2004). Neural correlates of first person perspective as one constituent of human self-consciousness. *Journal of Cognitive Neuroscience, 16*(5), 817–827.
Vuilleumier, P. (2009). The role of the human amygdala in perception and attention. In P. J. Whalen & E. A. Phelps (Eds.), *The human amygdala* (pp. 220–249). New York, NY: The Guilford Press.
Vygotsky, L. (1978). *Mind in society: The development of higher psychological processes.* Cambridge, MA: Harvard University Press.
Wachholtz, A. B., & Pargament, K. I. (2005). Is spirituality a critical ingredient of meditation? Comparing the effects of spiritual meditation, secular meditation, and relaxation on spiritual, psychological, cardiac, and pain outcomes. *Journal of Behavioral Medicine, 28*(4), 369–384.
Waldrop, M. M. (1992). *Complexity: The emerging science at the edge of order and chaos.* New York, NY: Simon & Schuster.
Walkerdine, V. (1984). Developmental psychology and the child-centered pedagogy: The insertion of Piaget into early education. In J. Henriques, W. Hollway, C. Urwin, C. Venn, & V. Walkerdine (Eds.), *Changing the subject: Psychology, social regulation and subjectivity* (pp. 153–202). UK, Methuen; New York, NY: Routledge.

Walkerdine, V. (1997). Redefining the subject in situated cognition theory. In D. Kirshner & J. Whitson (Eds.), *Situated cognition: Social, semiotic and psychological perspectives* (pp. 57–70). Mahwah, NJ: Lawrence Erlbaum Associates.
Wallace, B. A. (1999). The Buddhist tradition of Samatha. Methods of refining and examining consciousness. In F. Varela, & J. Shear (Eds.), *The view from within. First-person approaches to the study of consciousness* (pp. 175–187). Thorverton, UK: Imprint Academic.
Wang, D. J., Rao, H., Korczykowski, M., Wintering, N., Pluta, J., Khalsa, D. S., & Newberg, A. B. (2011). Cerebral blood flow changes associated with different meditation practices and perceived depth of meditation. *Psychiatry Research, 191*(1), 60–67.
Wangual, T. (1993). *Wonders of the natural mind. The essence of Dozgchen in the native Bon tradition of Tibet.* Barryton, NY: Station Hill.
Watts, A. (1975). *Psychotherapy east and west.* New York, NY: Vintage Books.
Weeks, E. (1966). *The Lowells and their institute.* Boston, MA: Little, Brown and Co.
Wenk-Sormaz, H. (2005). Meditation can reduce habitual responding. *Alternative Therapies in Health and Medicine, 11*(2), 42–58.
Wertz, F. J. (1983). Revolution in psychology: Case study of the new look school of perception. In A. Giorgi, A. Barton, & C. Maes (Eds.), *Duquesne studies in phenomenological psychology.* Pittsburgh, PA: Duquesne University Press.
Wertz, F. (1993). Cognitive psychology: A phenomenological critique. *Journal of Theoretical & Philosophical Psychology, 13*(1), 1–24.
Wertz, F. J. (Ed.). (1994). *The humanistic movement: Recovering the person in psychology.* New York, NY: Gardner Press.
Whalen, P. J., Kagan, J., Cook, R. G., Davis, F. C., Kim, H, & Polis, S.,…Johnstone, T. (2004). Human amygdala responsivity to masked fearful eye whites. *Science, 306,* 2061.
Wiener, N. (1948). *Cybernetics of control and communication in the animal and in the machine.* New York, NY: Wiley.
Wildman, W. J., & Brothers, L. A. (1999). A neurophysical-semiotic model of religious experience. In R. Russel, N. Murphy, T. Meyering, & M. Arbib (Eds.), *Neuroscience and the person. Scientific perspectives on divine action* (pp. 347–417). Berkeley, CA: Center for Theology and Natural Science.
Wilson, E. A. (1998). *Neural geographies.* London, UK: Routledge.
Wink, J. (2010). *Critical pedagogy: Notes from the real world.* New York, NY: Prentice Hall.
Winograd, T., & Flores, F. (1987). *Understanding computers and cognition: A new foundation for design.* Indianapolis, IN: Addison-Wesley Professional.
Wittig, A. F. (2001). *Introduction to psychology.* New York, NY: McGraw-Hill.
Woods, J. H. (Trans.) (1917). *The yoga sutras of Patanjali. Harvard Oriental Series, 17,* Cambridge, MA: Harvard University Press.
Xiong, G. L., & Doraiswamy, P. M. (2009). Does meditation enhance cognition and brain plasticity? *Annals of the New York Academy of Sciences, 1172,* 63–69.
Yakhlef, A. (2010). The corporeality of practice-based learning. *Organization Studies, 31*(4), 409–430.
Zahavi, D. (2001). Beyond empathy: Phenomenological approaches to intersubjectivity, *Journal of Consciousness studies, 8*(5–7), 151–167.
Zahavi, D. (2002). Anonymity and first-personal givenness: An attempt at reconciliation. In D. Carr & C. Lotz (Eds.), *Subjektivität – Verantwortung – Wahrheit* (pp. 75–89). Frankfurt am Main, Germany: Peter Lang.
Zahavi, D. (2003). *Husserl's phenomenology.* Stanford, NY: Stanford University Press.
Zahavi, D. (2005). *Subjectivity and selfhood: Investigating the first-person perspective.* Cambridge, MA: The MIT Press.
Zahavi, D. (2008). *Subjectivity and selfhood: Investigating the first-person perspective.* Cambridge, MA: The MIT Press.

Zahavi, D. (2010). Naturalized phenomenology. In S. Gallagher & D. Schmicking (Eds.), *Handbook of phenomenology and cognitive science* (pp. 3–19). Dordrecht, UK, New York, NY: Springer.

Zahavi, D., & Parnas, J. (1999). Phenomenal consciousness and self-awareness: A phenomenological critique of representational theory. In S. Gallagher & L. Shear (Eds.), *Models of the self* (pp. 253–270). Exeter, UK: Imprint Academic.

Zahavi, D., & Roepstorff, A. (2011). Faces and ascriptions: Mapping measures of the self. *Consciousness and Cognition, 20*, 141–148.

Zanto, T. P., Rubens, M. T., Thangavel, A., & Gazzaley, A. (2011). Causal role of the prefrontal cortex in top-down modulation of visual processing and working memory. *Nature Neuroscience, 14,* 656 – 661.

Zeman, A. (2001). Consciousness. *Brain, 124*(7), 1263–89.

Zull, J. E. (2002). *The art of changing the brain: Enriching teaching by exploring the biology of learning.* Sterling, VA: Stylus Publishing.

Index

A
Adrenocorticotropic hormone (ACTH), 133, 134, 141
Age of Theory, 108
Autopoiesis
  neurophenomenological praxis, 39
  psychoneurointracrinology
    cybernetics/dynamic systems theory, 117
    enactive approach, 117
    structural determinism, 117
    theory, 117 (*see also* Psychoneurointracrine autopoiesis theory)
    Varela (*see* Varela)

B
Blooming, buzzing confusion, 107

C
Central nervous system (CNS), 132–134, 136
Circumventricular organs (CVOs), 132, 134
Clark Hull's hypothetico-deductive method, 108
Cognitive paradox, 106
Cognitive phenomenology
  analysis
    dynamic systemic restructure, 77
    isolation of, 77–78
    vs. meditation styles, 74–76
    neural network, 77
    switching off process, 77
    Vedantic and Hesychastic types, 77
    visual modality, 76
  brain-imaging technology, 61
  experience gap, 62
  great intensity variable, 68
  heightened intuition/insight, 62
  mapping (*see* Phenomenological-cognitive mapping)
  meditation and scientific research
    alpha-blocking activity, 65
    analytic paradigms, 65
    EEG researcher, 65
    intrinsic brain state, 64
    physiological and psychological effect, 64
    simple relaxation technique, 64
    systematic reflective analysis, 65
    theoretical and mathematical model, 64
    theoretical framework, 65
  meditation styles, 66–67
  mutual enlightenment, 62
  psychology, history of, 67
  repetitive and reproducible quality, 62
  visual imagery, 63
  visual-spatial processing, 61
  VWM function, 63

D
Darwin's theory, 91
Dehydroepiandrosterone (DHEA), 130–132, 135, 136

E
Emotion theory, 124–125
Endocrinology, 116
Epoché method
  autoethnography, 36–37
  human perception, 36
  natural attitude, 35–36

DOI 10.1007/978-1-4614-7239-1,

Epoché method (*cont.*)
  object/subject dualis, 35
  systematic and thorough analysis, 37
Exceptional mental states, 90, 99–100, 106

G

γ-aminobutyric acid (GABA), 125, 131–133, 142–143
Gonadotropin-releasing hormone (GnRH), 134, 135

H

Heidegger's theory
  Befindlichkeit, 18
  brain, 20–21
  cognitive science, 19–20
  ontological category, 18
  present-at-hand mode, 17
  ready-to-hand mode, 17
  world, 17–18
Human menstrual cycle, 116, 135–136
Husserl's phenomenological method, 69
Hypothalamic-pituitary-adrenal (HPA), 115, 131–135
Hypothalamic-pituitary-gonadal (HPG), 115, 129, 132–134

I

Ignis fatuus, 96

J

James, William
  blooming, buzzing confusion, 107
  The Brain and the Mind
    brain neurophysiology, 91
    brutal materialism lurking, 92
    cellular structure, 93
    cerebral hemisphere, 93
    contemporary theory, 95
    Darwin's theory, 91
    evolutionary scale, 93
    fictitious anatomy, 93
    ignis fatuus, 96
    metaphysical club, 91
    obsolete method, 93
    philosophical implication, 92
    phrenology presumes, 94
    physiological theory, 94
    summum bonum model, 95
    textile mill development, 90
  brain process, 106
  cognitive paradox, 106
  consciousness, 100–102
  evolutionary process, 97
  exceptional mental states, 90, 99–100
  hypnosis and *idea fixe*, 106
  vs. modern neurophenomenologist, 103–105
  neuroscience revolution, 89–90
  pragmatism, 90
  psychology
    Age of Theory, 108
    Clark Hull's hypothetico-deductive method, 108
    clinical psychology, 108
    cognitive-behavioral era, 110
    commonalities, 109
    definition, 111
    implication of, 107
    neuroscience revolution, 110
    neurotheology, 109
    reductionistic materialism, 110
    Western scientific model, 109
  radical empiricism, 98–99, 106
  rationality, 96
  self-governing system, 89
  the specious present, 97–98, 105
  tripartite metaphysics, 102–103

L

LeDoux's model, 5

M

Mandala and Deity Yoga, 70, 71
Monoamine oxidase (MAO), 132, 135–136

N

Neurohormonal mechanism, 133
Neuromodulatory mechanism, 133
Neurophenomenological praxis
  affordances, 40
  attractors, 40
  autopoietic system, 39
  brain, 47–48
  brain-based learning, 27
  cognitive neuroscience, 33
  cognitive psychology, 30–32
  cognitivism, 29–30
  complexity, 39
  distributed activity, 41
  embodied cognition, 25

enactive cognition, 26
learning, 27–29
method
aspect of, 37
eidetic reduction, 38
epoché, 35–37
fMRI and PET scans, 38
invariant and essential structures, 38
prosaic/hackneyed, 37
sensory descriptions, 37
Wertz's contention, 38, 39
methodology, 34–35
neurophenomenology, 26
operational closure, 39–40
pedagogical praxis
game identity, 45–46
sabotage, 43–44
saliva/spit, 46–47
social construction, 42–43
pedagogy, 41–42
perturbations, 40
phenomenology, 33
portfolio assessment, 26–27, 49
praxis, 26
self-organization, 39
semiotic trajectory, 41
situated learning, 25
structural coupling, 40
trajectory, 41
Neurophenomenology emotion
affective neuroscience
cognitive functions, 6
emotional experience, 4
experience, 6
implication, 6
LeDoux's model, 5
mammillary body, 5
neocortex, 5
brain's limbic system, 2
cognitive neuroscience, 1
cognitive revolution
affective revolution, 2, 3
AI research, 4
assumptions, 3
experience, 4
human context, 3
interpretation process, 2
mentalistic language, 2
eidetic reduction, 8
emotional experience, 2
enactive approach, 16–17
folk psychology, 8
Heidegger's theory (*see* Heidegger's theory)
neurophilosophy, 7
phenomenological psychology, 1
phenomenology, 7–8
physiological fact, 10
quantitative method, 9
reductionism
comportment, 10
dialectical relation, 11
explanatory gap, 10
human order, 12–14
implication, 11–12
naturalism, 14, 15
phenomenological attitude, 14–16
physical order, 12
physical stimuli, 11
structure, 10
vital order, 12
social context, 9
Neuroscience revolution, 89, 90, 108, 110
Neurotheology, 109
Newtonian, Kantian, and Aristotelian model, 111
N-methyl-D-aspartate (NMDA), 132

P

Paraventricular nucleus (PVN), 133–134
Pedagogical praxis
game identity, 45–46
sabotage, 43–44
saliva/spit, 46–47
social construction, 42–43
Phenomenological analysis
dynamic systemic restructure, 77
isolation of, 77–78
vs. meditation styles, 74–76
neural network, 77
switching off process, 77
Vedantic and Hesychastic types, 77
visual modality, 76
Phenomenological-cognitive mapping
cognitive psychological variable, 72–73
experimental design, 73–74
hermeneutic cycle, 69
Husserl's phenomenological method, 69
meditation training, 70
meditative *vs.* non-mediative mode, 71–72
referential/non-referential mode, 70
transformation, 70–71
Pragmatism, 90, 91, 100, 102, 103, 107
Psychology
affective revolution, 2–3
cognitive science, 19–20
James, William
Age of Theory, 108

Psychology (*cont.*)
Clark Hull's hypothetico-deductive method, 108
clinical psychology, 108
cognitive-behavioral era, 110
commonalities, 109
definition, 111
implication of, 107
neuroscience revolution, 110
neurotheology, 109
reductionistic materialism, 110
Western scientific model, 109
natural attitude, 7
Psychoneurointracrine autopoiesis theory
allosteric enzymatic regulation, 129
circumventricular organ, 132
CNS function, 132
dehydroepiandrosterone, 130–131
genomic/non-genomic pathway, 131, 132
HPG/HPA axes, 129, 132
MAO activity, 132
neurohormonal/neuromodulatory mechanism, 133
reactive-anticipatory, 131
steroid biological activity, 129
steroid synthesis, 129–130
stress integration mechanism, 130–131
up and down regulation, 129
Psychoneurointracrinology
autopoiesis
cybernetics/dynamic systems theory, 117
enactive approach, 117
structural determinism, 117
theory, 117 (*see also* Psychoneurointracrine autopoiesis theory)
Varela (*see* Varela)
CRF-ACTH pathway, 116
definition, 115
embodied sense of self
amygdala, 125–127
brain process, 124
cognitive research variable, 123
emotion theory, 124–125
ephemeral theoretical entity, 123
fluctuation/qualified state, 123
mythopoetic function, 123
subliminal self, 123
embodiment
body image/schema, 122–123
definition, 122
extended body theory, 122
indisciplinary approach, 122
intentional object/goal, 121
newborn' s primitive body image, 122
proprioceptive self, 122
role of, 120
sensation, 121
sense of ownership, 120
enactive cognition, 115
endocrinology, 116
existential-phenomenological state, 116
growth-oriented transcendent dimension
autopoiesis theory (*see* Psychoneurointracrine autopoiesis theory)
center of gravity, 128
emergent global states theory, 133–134
existential moment, 127
human development, 128
metaphysics, 127
mythopoetic function, 129
somatic process, 129
HPG/HPA axes, 115
human menstrual cycle, 135–136
interdisciplinary approach, 116
prereflectivity
body image/schema, 121
consciousness, 121
dynamic sensorimotor principles, 121
structure of, 120

R
Radical empiricism
James, William
definitions of, 101
exceptional mental states, 106
functionalism, 110
Principles of Psychology, 98–99, 107
tripartite metaphysics, 90, 100, 102
Rig-pa awareness, 70, 74

S
Summum bonum model, 95

T
The Brain and the Mind
brain neurophysiology, 91
brutal materialism lurking, 92
cellular structure, 93
cerebral hemisphere, 93
contemporary theory, 95
Darwin's theory, 91
evolutionary scale, 93
fictitious anatomy, 93

ignis fatuus, 96
metaphysical club, 91
obsolete method, 93
philosophical implication, 92
phrenology presumes, 94
physiological theory, 94
summum bonum model, 95
textile mill development, 90

V
Varela
autopoietic system, 116
biochemical pathway, 118
immune system, 119
nervous and lymphoid system, 119
non-embodied approach, 118
ongoing pattern, 120
organism identity, 120
phenomenal experience/embodiment, 118
recognition event, 119
"sense of self,",118, 119, 123
sensorimotor system, 118
Vedantic and Hesychastic types, 77
Visual working memory (VWM), 63, 74

W
Wertz's contention, 38, 39
Western scientific model, 109–110

MIX
Papier aus verantwortungsvollen Quellen
Paper from responsible sources
FSC® C105338

If you have any concerns about our products,
you can contact us on
ProductSafety@springernature.com

In case Publisher is established outside the EU,
the EU authorized representative is:
Springer Nature Customer Service Center GmbH
Europaplatz 3, 69115 Heidelberg, Germany

Printed by Libri Plureos GmbH
in Hamburg, Germany